Hardwood Floors

2nd Edition

Dan Ramsey

TAB BOOKS
Blue Ridge Summit, PA

Therefore everyone who hears these words of
mine and puts them into practice is like a wise
man who built his house on the rock. The rain
came down, the streams rose, and the winds
blew and beat against the house; yet it did not
fall, because it had its foundation on the rock.

Matthew 7:24-25 (NIV)

SECOND EDITION
SECOND PRINTING

© 1991 by **TAB Books**.
TAB Books is a division of McGraw-Hill, Inc.

Library of Congress Cataloging-in-Publication Data

Ramsey, Dan, 1945-
    Hardwood floors / by Dan Ramsey. — 2nd ed.
        p.   cm.
    Includes index.
    ISBN 0-8306-7529-9   ISBN 0-8306-3529-7 (pbk.)
    1. Floors, Wooden. 2. Hardwoods. I. Title.
TH2529.W6R36   1990
694'.2—dc20   90-49718
                                                CIP

TAB Books offers software for sale. For information and a catalog, please contact
TAB Software Department, Blue Ridge Summit, PA 17294-0850.

Acquisitions Editor: Kimberly Tabor
Book Editor: Barbara Minich
Production: Katherine G. Brown
Book Design: Jaclyn J. Boone
Cover Design: Lori E. Schlosser

Contents

Acknowledgments

*M*any professionals have contributed their knowledge and skills to the completeness of the first and second editions of this book. They should be thanked: National Oak Flooring Manufacturers Association; Maple Flooring Manufacturers Association; American Parquet Association, Inc.; Wood and Synthetic Flooring Institute; Harris-Tarkett, Inc; Lavidge & Associates, Inc.; North American Hardwood Flooring; The Jennison-Wright Corp.; Dixon Lumber Company, Inc.; Oregon Lumber Company; Applied Radiant Energy Corp.; Memphis Hardwood Flooring Co.; Oregon Lumber Co.; Horner Flooring Co.; Zickgraf Hardwood Co.; Robbins/Sykes; Hardwood floor refinisher, Jerry Spicer; and U.S. Department of Agriculture, Forest Service. Thank you!

A special thanks goes to Heather Ramsey for typing the manuscript and preparing the artwork for this second edition.

Introduction

*H*ardwood floors, which were popular a few decades ago, are returning to new and remodeled homes for both fashion and function. Meanwhile, older homes are being restored and seasoned hardwood floors are being reconditioned.

Hardwood Floors—2nd Edition introduces the benefits of wood flooring to the new consumer. It thoroughly covers the planning, selection, installation, finishing, maintenance, repair, and refinishing of all types of hardwood flooring. Tongue-and-groove strip flooring is the most popular. Also covered are plank, block, and parquet floors. Step-by-step instructions and illustrations show you how to tackle every aspect of a hardwood floor installation and repair like a professional.

Chapter 1 clearly explains how to decorate or remodel your home by installing or refinishing your hardwood floor. Chapter 2 guides you in the selection of the tools and materials you'll need for the job—plus tips on how to hire the right flooring contractor if you choose to do so.

Chapter 3 offers clear instruction on how to install your hardwood floor no matter what the condition. You'll learn the special techniques required for plank, tongue-and-groove, parquet, and block hardwood flooring.

One of the greatest benefits of hardwood floors is their natural wood beauty. Chapter 4 shows you how to finish your hardwood floor to match your decor by using oak, maple, or other hardwoods plus the newer special finishes. Clear photos illustrate the refinishing process that made a 50 year-old hardwood floor look like new.

If you have a new or existing hardwood floor, Chapter 5 tells you how to maintain it for many years of easy care. It answers common questions about the new finishes that are available to maintain them as good as new. You'll also learn how to perform simple repairs on your hardwood floor and its finish.

Chapter 6 is written especially for those who have older hardwood floors that need to be refinished. It clearly explains how to select the tools and materials you will need as well as how to completely sand, stain, and refinish any hardwood floor. It also shows you how to repair common problems before you refinish the floor.

To make this book complete, *Hardwood Floors—2nd Edition* includes two appendices: hardwood floor data tables and hardwood floor game markings.

A best-seller in its first edition, *Hardwood Floors—2nd Edition* offers the latest materials and techniques for making your hardwood floors beautiful and easy to maintain for generations.

Chapter **1**

Understanding hardwood floors

Wood flooring has been a traditional American favorite since pioneer homes were first erected from that most readily available resource—the tree. Today, however, hardwood flooring is expensive. The selection and installation of quality hardwood flooring often costs more than other common flooring materials. Even so, the lower maintenance costs and the aesthetic qualities of hardwood flooring, combined with reduced costs to the do-it-yourselfer, have spurred a renewed interest in practical hardwood floors. An important reason is the natural beauty of hardwood floors (FIGS. 1-1 through 1-2), which blends with all types and tastes of interior decorating from traditional to contemporary.

HARDWOOD FLOORING BASICS

Wood flooring materials are available in strip, plank, block, and tile form. Wood flooring is available in both hardwood and softwood. The most common types of hardwood floors are red and white oak, beech, birch, maple, and pecan. The softwoods most commonly used for flooring are southern pine, Douglas fir, redwood, and western hemlock.

Softwood finish flooring costs less than most hardwood species and can be used where traffic is light. However, softwood flooring is less dense than the hardwoods so it is less wear-resistant and more readily shows abrasions. Softwood finish flooring is also less expensive than hardwood flooring.

The most popular hardwood flooring is $25/32 \times 21/4$-inch strip flooring. The strips are laid lengthwise in a room and normally at right angles to the floor joists. Strip flooring is typically tongue-and-groove and end-matched. Strips are of random length and may vary from 2 to 16 feet or longer.

I-I Hardwood floors make a kitchen or dining room more traditional (courtesy Robbins/Sykes).

End-matched strip flooring is usually hollow-backed. The face is wider than the back so that when it is assembled on a floor the joints are tight. A tight fit means fewer floor squeaks.

Wood block flooring uses tongued-and-grooved blocks of wood that measure $25/32$ (about $3/4$) inch thick and are available in dimensions from 4×4 to 9×9 inches. Larger measurements are preferred for larger rooms to seemingly reduce their size.

I-2 Parquet hardwood floors fit well in more contemporary homes (courtesy Robbins/Sykes).

PROPERTIES OF WOOD

To guide you to an understanding of hardwood flooring (FIG. 1-3), let's consider the properties of wood, the types of wood joints, the common types of hardwood flooring, and how they all fit into modern home construction.

Any piece of wood is made up of a number of small cells as shown in FIG. 1-4. The size and arrangement of the cells determine the grain of the

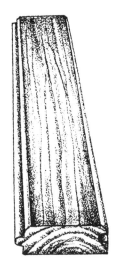

1-3 Typical hardwood flooring strip
(courtesy National Oak Floor
Manufacturers Association).

wood and many of its properties. Examine a freshly cut tree stump and you'll see that the millions of small cells are arranged in circular rings around the pith, or center, of the tree (FIG. 1-5). These rings are caused by a difference in the rate of growth of the tree during the various seasons of the year. In spring, the tree grows rapidly and builds up a thick layer of comparatively soft, large cells that appear in the cross section of the trunk as the light-colored annual rings.

As the weather becomes warmer during the early summer, the rate of growth slows. The summer growth forms cells that are packed more closely. These pairs of concentric springwood and summerwood rows form the annual rings, which can be counted to find out the age of the

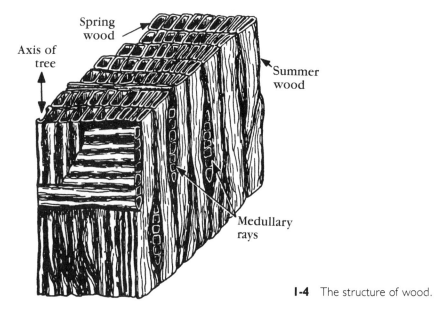

1-4 The structure of wood.

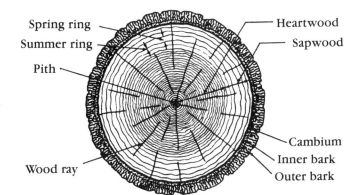

Spring ring — Heartwood
Summer ring — Sapwood
Pith —
Cambium
Inner bark
Outer bark
Wood ray

1-5 Cross section of a tree.

tree. Because of climatic conditions, some trees, such as oak and walnut, have more distinctive rings than others, such as maple and birch. White pine is so uniform that you can hardly distinguish the rings, while many other softwoods have a very pronounced contrast between summerwood and springwood, which makes it easy to distinguish the rings.

The sapwood of a tree is the outer section of the tree between the heartwood, or the darker center wood, and the bark. The sapwood is lighter in color than the heartwood, but it gradually becomes darker as it changes to heartwood on the inside and as new layers are formed on the outside. Depending upon the type of tree, it requires from 9 to 36 years to transform sapwood into heartwood.

The cambium layer, which is the boundary between the sapwood and the bark, is the thin layer where new sapwood cells form. Medullary rays are radial lines of wood cells that consist of threads of pith and serve as the lines of communication between the central cylinder of the tree and the cambium layer. They are especially prominent in oak.

When a tree is sawed lengthwise, the annual rings form a pattern, which is called the grain of the wood. Many terms are used to describe the various grain conditions. If the cells that form the grain are closely packed and small, the wood is said to be fine-grained or close-grained. Maple and birch are excellent examples of this type of wood. If the cells are large, open, and porous, the wood is coarse-grained or open-grained. Examples of this are oak, walnut, and mahogany. Furniture and flooring made of open-grained woods require the use of a wood filler to close the pores and provide a smooth outside finish.

When the wood cells and fibers are comparatively straight and parallel to the trunk of the tree, the wood is said to be straight-grained. If the grain is crooked, slanted, or twisted, it is said to be cross-grained. It is the arrangement, direction, size, and color of the wood cells that give the grain of each wood its characteristic appearance.

CUTTING AND SEASONING LUMBER

In large lumber mills, such as those found in the Pacific Northwest, logs are usually processed into lumber with huge band or circular saws. There

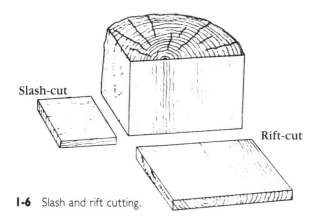

1-6 Slash and rift cutting.

are two methods of sawing up logs. Slash cutting is accomplished by a series of cuts that are made parallel to the side of the log. If hardwoods are being cut, the process is known as plain sawing. If softwoods are being cut, the process is referred to as flat-grain sawing.

Lumber that is specifically cut to provide edge grain on both faces is said to be rift-cut (FIG. 1-6). If hardwood is being cut, the lumber is said to be quartersawed. If softwood is being sawed, it is called edge-grain lumber. Incidentally, if an entire log is slash-cut, several boards from near the center of the log will actually be rift-cut.

Slash-cut lumber is usually cheaper than rift-cut lumber because it takes less time to slash-cut a log and there is less waste. Circular or oval knots that appear in slash-cut boards affect the strength and surface appearance much less than do spike knots, which may appear in rift-cut boards. If a log is sawed to produce all slash-cut lumber, however, more boards will contain knots than if the log were sawed to produce the maximum amount of rift-cut material. Another advantage of slash cutting is that when shakes and pitch pockets are present, they will extend through fewer boards.

For many applications, especially flooring, rift-cut lumber is preferred because it is more wear-resistance than slash-cut lumber. Rift-cut lumber also shrinks and swells less in width. Another advantage to rift-cut lumber over slash-cut lumber is that it twists and cups less and splits less when used. Rift-cut lumber also usually holds paint and other finishes better.

After being sawed, lumber must be thoroughly dried before it is suitable for most uses. The old method—and one still preferred for some uses—was merely to air-dry the lumber in a shed or stack it out in the open. This method requires considerable time for the wood to dry—up to seven years for some of the hardwoods.

A faster method is known as kiln drying. The wood is placed in a tight enclosure, called the kiln, and dried with heat that is supplied by artificial means. The length of drying time required varies from two or three days to several weeks, depending on the kind of wood, its dimensions, and the method of drying.

Lumber is considered dry enough for most uses when the moisture content has been reduced to between 12 and 15 percent. If you use lumber very much, you will soon learn to judge the dryness of wood by its color, weight, smell, feel, and by a visual examination of shavings and chips. Your lumber supplier can also give you a close estimate of a wood's dryness.

Briefly, lumber is seasoned by removing the moisture from the millions of small and large cells of which wood is composed. Moisture, which can be water or sap, occurs in two separate forms: free water and embedded water. Free water is the amount of moisture the individual cells contain. Embedded water is the moisture absorbed by the cell walls.

During the drying or seasoning process, the free water in the individual cells evaporates until a minimum amount of moisture is left. The point at which this minimum moisture remains is called the fiber saturation point. The moisture content at this point varies from 25 to 30 percent. Below the fiber saturation point, the embedded water is extracted from the porous cell walls. This process causes a reduction in the thickness of the walls. Wood shrinks across the grain when the moisture content is lowered below the fiber saturation point.

Wood shrinks or swells when varying amounts of moisture change the size of the cells. Therefore, the lowering or raising of the moisture content causes lumber to shrink or swell. The loss of moisture during the seasoning process causes wood to be harder, stronger, stiffer, and lighter in weight—all qualities important to the look of hardwood flooring.

WOOD FLOORING MATERIALS

There is a wide selection of wood materials that may be used for flooring. Hardwoods and softwoods are available as strip flooring (FIG. 1-7) in a variety of widths and thicknesses, and as random-width planks, and block flooring (FIG. 1-8).

Softwood finish flooring costs less than most hardwood species and is often used to good advantage in bedroom and closet areas where traffic is light. It might also be selected to fit the interior decor. While this book primarily covers hardwood flooring, many of the principles and instructions also apply to softwood flooring.

TABLE 1-1 lists the grades and description of softwood strip flooring. Softwood flooring has tongue-and-groove edges and may be hollow-backed or grooved. Some types are also end-matched. Vertical-grain flooring generally has better wearing qualities than flat-grain flooring. TABLE 1-1 also lists the grades, types, and sizes of hardwood strip flooring. Manufacturers supply both prefinished and unfinished flooring.

Perhaps the most widely used flooring pattern is the $25/32 \times 21/4$-inch strip flooring. The strips are laid lengthwise in a room and normally at right angles to the floor joists. A subfloor of diagonal boards or plywood is normally used under the finished floor. Strip flooring of this type is tongue-and-groove and end-matched (FIG. 1-9). Strips are of random length and may vary from 2 to 16 feet or more.

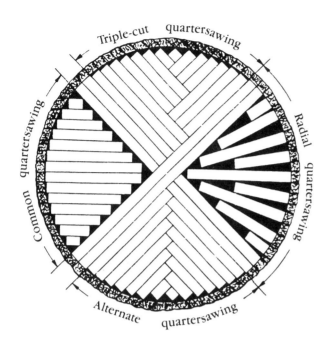

1-7   Four methods of quartersawing.

End-matched strip flooring in a $^{25}/_{32}$-inch thickness is generally hollow-backed. The face is slightly wider than the bottom so that tight joints result when the flooring is laid. The tongue fits snugly into the groove to prevent movement and floor squeaks. All of these details are designed to provide beautiful finished floors that require a minimum of maintenance.

Another matched pattern may be obtained in $^3/_8 \times 22$-inch strips (FIG. 1-10). It is commonly used for remodeling work or when the subfloor is edge-blocked or thick enough so that there is very little defection under heavy loads.

Square-edged strip flooring (FIG. 1-11) might also be used occasionally. It is usually $^3/_8 \times 2$ inches in size and is laid up over a substantial subfloor. Face nailing is required for this type.

Wood-block flooring (FIG. 1-12) is made in a number of patterns. Blocks may vary in size from 4×4 to 9×9 inches and larger. The thickness varies by type from $^{25}/_{32}$ inch for laminated blocking or plywood block tile to $^1/_8$-inch stabilized veneer. Solid wood tile is often made up of nar-

1-8   Strip flooring is available in vertical (A) and flat-grain (B) hardwood and softwood stock.

Table 1-1 Wood Flooring Grades and Graining

Species	Grain orientation	Size		First grade	Second grade	Third grade
		Thickness, inches	*Width, inches*			
		Softwoods				
Douglas-fir and hemlock	Edge grain	3/4	1 1/8 − 5 1/8	B and Better	C	D
	Flat grain	3/4	1 1/8 − 5 1/8			
Southern pine	Edge grain and Flat grain	5/16 − 1 1/4	1 1/8 − 5 1/8	B and Better	C, C and Better	D (and No. 2)
		Hardwoods				
Oak	Edge grain	3/4	1 1/2 − 3 1/4	Clear	Select	------------------------
	Flat grain	11/32	1 1/2, 2 ⎫		Select	No. 1 Common
		15/32	1 1/2, 2 ⎭	Clear		
Beech, birch, maple, and pecan[1]		25/32	1 1/2 − 3 1/4 ⎫			
		3/8	1 1/2, 2 1/4 ⎬	First grade	Second grade	Third grade
		1/2	1 1/2, 2 1/4 ⎭			

[1]Special grades are available in which uniformity of color is a requirement.

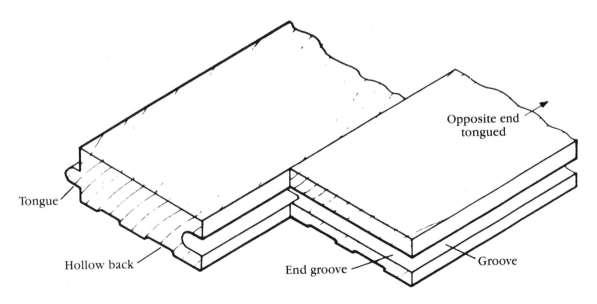

1-9 Tongue-and-groove, end-matched strip flooring.

Tongue

Hollow back

Opposite end tongued

End groove

Groove

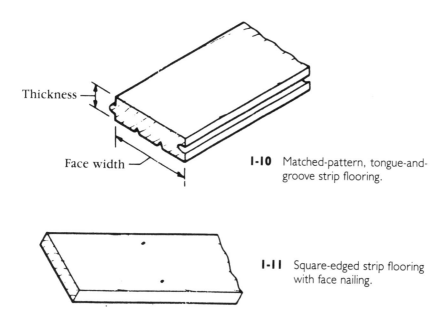

Thickness

Face width

1-10   Matched-pattern, tongue-and-groove strip flooring.

1-11   Square-edged strip flooring with face nailing.

row strips of wood that have been splined or keyed together in a number of ways. Edges of the thicker tile are tongue-and-groove, but thinner sections of wood are usually square-edged (FIG. 1-13). Plywood blocks may be ³/₈ inch or thicker and are usually tongue-and-groove (FIG. 1-14). Many block floors are factory-finished and need only be waxed after installation (FIG. 1-15).

Figure 1-16 illustrates a typical installation of a tongue-and-groove parquet-block floor unit. Figure 1-17 illustrates cross matching parquet tiles.

1-12   Wood-block flooring.

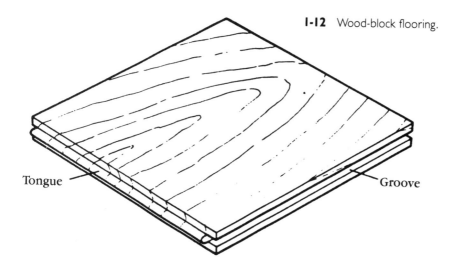

Tongue

Groove

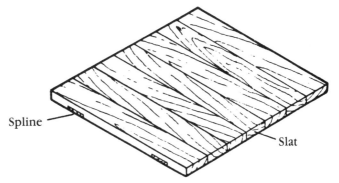

1-13   Square-edged and splined wood-block flooring.

1-14   Tongue-and-groove laminated block flooring.

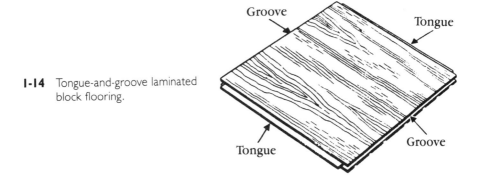

WOOD JOINTS

Another important element of hardwood flooring is the way in which the wood strips or blocks are joined. I've already mentioned the tongue-and-groove joint. Other types of joints that are used in woodworking include the butt, lap, miter, rabbet, dado, gain, mortise-and-tenon, slip tenon, box corner, and dovetail. Few are ever used, however, for the manufacturing and installation of hardwood floors.

Lap joints (FIG. 1-18) can be used for hardwood flooring, but are usually not. The more common types of lap joints are the plain lap, cross half-lap, end butt half-lap, and corner half-lap.

Figure 1-19 illustrates the simplest of all joints—the plain butt joint. It is made by butting or placing two pieces of wood together. As mentioned earlier, butt joints must be face-nailed, or nailed through the face of the wood, in order to be fastened securely.

Figure 1-20 shows one way of overcoming this problem by using a doweled joint. Holes are drilled and connecting dowels are installed in order to fasten the members together in this joint.

Taking the doweled joint one step further is the spline joint (FIG. 1-21). A notch is cut in both members and a smaller member, which is called a spline, is placed between them. The milling and insertion of a spline joint takes time and slows down the installation of the flooring.

I-15   Factory-finished wood-block flooring (courtesy Pennwood Products Co.).

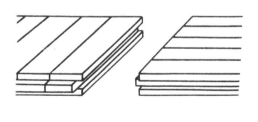

I-16   Tongue-and-groove parquet-block floor unit (courtesy Pennwood Products Co.).

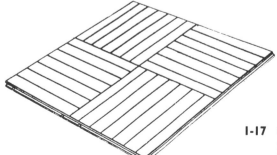

I-17   Cross-matched parquet wood tiles (courtesy Pennwood Products Co.).

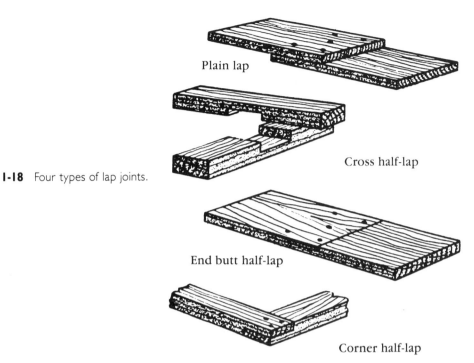

Plain lap

Cross half-lap

1-18   Four types of lap joints.

End butt half-lap

Corner half-lap

The most popular joint for flooring is called the tongue-and-groove (FIG. 1-22). In this joint, the spline is actually milled from part of one of the wood members. The spline is then called a tongue and is designed to fit into a groove—hence the name. This type of joint offers a rigid connection between the flooring pieces and does not require an amount of time for assembly.

Figure 1-23 illustrates one popular type of floor joint that is actually a combination of joints. As you can see, the sides are butt-jointed. The edge

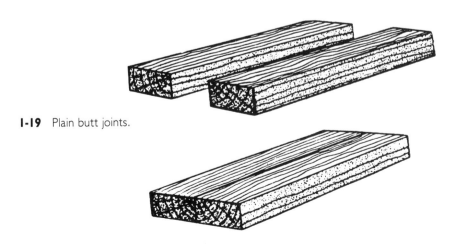

1-19   Plain butt joints.

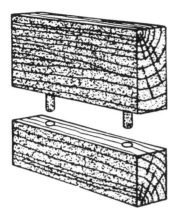

I-20   Doweled joint.

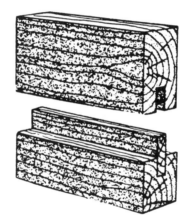

I-21   Spline joint.

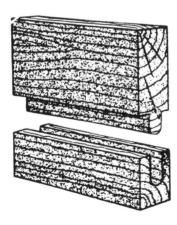

I-22   Tongue-and-groove joint.

1-23 Typical manufactured hardwood flooring joint (courtesy Maple Flooring Manufacturers Association

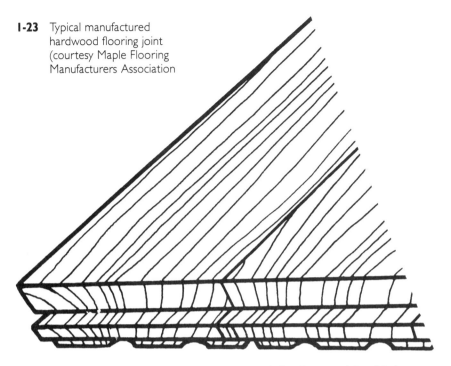

is a tongue-and-groove joint. If installed tightly, this combined joint can be quicker and just as tight as having a full tongue-and-groove joint on all four sides.

FLOORING PATTERNS

There are numerous ways and patterns you can use to install hardwood flooring that will make your floor unique as well as functional. Figure 1-24

1-24 Identical length strip flooring (courtesy Pennwood Products Co.).

I-25   Herringbone design hardwood floor (courtesy Pennwood Products Co.).

illustrates how strip flooring can be installed when the boards are of the same length rather than random lengths.

Figure 1-25 shows a diagonal cross or herringbone design that is easy to install. It can be of identical strip flooring or of patterned block flooring.

Figures 1-26 through 1-34 illustrate a variety of block flooring patterns that can be found in many flooring departments or stores. They offer an infinite number of combinations that can beautify your home while saving you the labor of assembling and installing such intricate patterns.

HOUSE CONSTRUCTION

Figure 1-35 shows an exploded view of the major parts of a single story, wood-frame house. The floor system, interior and exterior walls, and the roof are the major components of such a house. Houses with flat or low-sloped roofs are usually variations of these systems.

The illustration shows a floor system that is constructed over a crawl space. The supporting beams are fastened to treated posts that are embedded in soil or to masonry piers or a foundation. The floor joists are

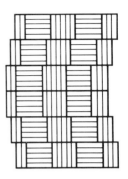

1-26 Staggered block flooring pattern (courtesy Harris-Tarkett, Inc.).

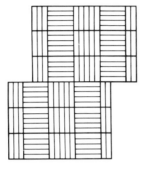

1-27 Overlap block flooring pattern (courtesy Harris-Tarkett, Inc.).

1-28 Monticello design block flooring (courtesy Pennwood Products Co.).

1-29 One-directional block flooring design (courtesy Pennwood Products Co.).

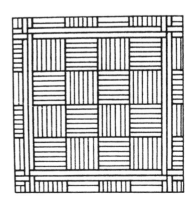

1-30 Log cabin design block flooring (courtesy Pennwood Products Co.).

1-31 Herringbone design block flooring (courtesy Pennwood Products Co.).

1-32 Ashlar design block flooring (courtesy Pennwood Products Co.).

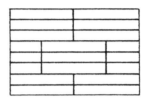

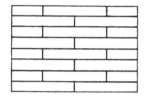

1-33 Brick design block flooring (courtesy Pennwood Products Co.).

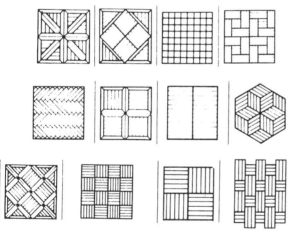

1-34 Assorted parquet hardwood flooring designs (courtesy Pennwood Products Co.).

fastened to these beams and the subfloor is nailed to the joists. This creates a level, sturdy platform upon which the rest of the house is constructed. This type of construction is often called platform construction.

Figure 1-36 shows the typical components of a hardwood floor. The joists support the subfloor. Building paper is laid down to prevent moisture from entering the flooring from below, which could possibly

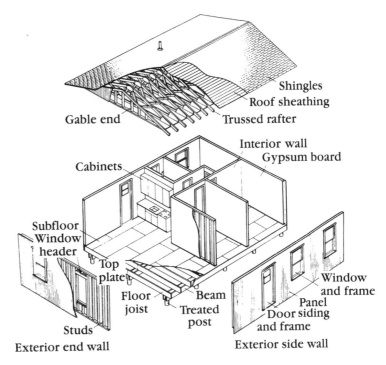

Shingles
Roof sheathing
Gable end
Trussed rafter

Interior wall
Gypsum board

Cabinets

Subfloor
Window
header
Top
plate
Floor
joist
Beam
Treated
post
Studs
Exterior end wall

Window
and frame
Panel
Door siding
and frame
Exterior side wall

1-35 Exploded view of a wood-frame home.

1-36 Laying out a hardwood floor over a subfloor.

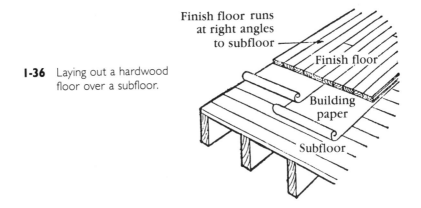

Finish floor runs
at right angles
to subfloor

Finish floor

Building
paper

Subfloor

damage the finish. Finally, finish flooring is installed, often at right angles to the subfloor.

Many of these construction terms will be new to you. The glossary at the back of the book defines these terms. Refer to it often; it will help you understand working with hardwood flooring.

Chapter **2**

Planning hardwood floors

A good job begins with good planning. This adage is especially true for installing and maintaining hardwood floors. Many installation problems can be solved on paper before the work begins, which saves time and materials. In this chapter, you'll learn how to plan and prepare for your hardwood flooring project by selecting the correct materials, properly choosing and using tools and fasteners, and learning how hardwood flooring can effectively be used in your home.

THE FLOOR PLAN

Figure 2-1 illustrates a floor or building plan. In home construction, the floor plan guides the contractor and subcontractors in the erection of walls, floors, and roof. Information found on a floor plan includes the lengths, thicknesses, and character of the building walls on each particular floor; the widths and locations of door and window openings; the lengths and character of partitions; the number and arrangement of rooms; and the types and locations of utility installations. In many cases, the floor plan will also establish the type and layout of the flooring.

A floor plan of your home can be valuable to you when you estimate, plan, and install hardwood flooring. You may have a floor plan for your home from the contractor or from an appraiser, or you may have to generate one yourself with a tape measure and some graph paper. In either case, you'll find the time spent on researching your home's floor plan a worthwhile investment as your hardwood flooring project continues.

PLANNING LUMBER

Hardwood flooring is lumber. You may also need to use other types of structural lumber as you install and repair your hardwood floor, such as

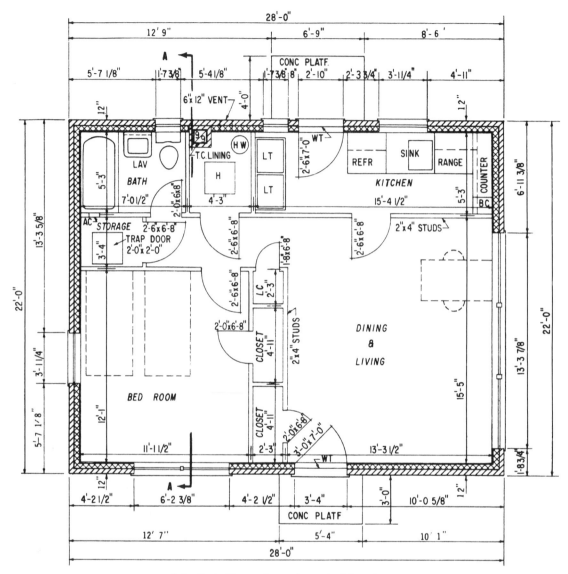

2-1 Typical home floor plan or building plan.

subflooring, floor joists, sole plates, wall studs, and headers. Let's consider the planning and selection of lumber.

Lumber is usually sawed into standard sizes, which are described in length, width, and thickness measurements. This permits uniformity when structures are planned and materials ordered. TABLE 2-1 lists the common widths and thicknesses of wood in rough, or nominal, and dressed dimensions in the United States. Standards have been established for dimension differences between the quoted size of the lumber and its

Table 2-1   Nominal and Standard Lumber Sizes

Nominal size(in.)	American standard (in.)
1 × 3	$25/32 \times 25/8$
1 × 4	$25/32 \times 35/8$
1 × 6	$25/32 \times 55/8$
1 × 8	$25/32 \times 71/2$
1 × 10	$25/32 \times 91/2$
1 × 12	$25/32 \times 111/2$
2 × 4	$15/8 \times 35/8$
2 × 6	$15/8 \times 55/8$
2 × 8	$15/8 \times 71/2$
2 × 10	$15/8 \times 91/2$
2 × 12	$15/8 \times 111/2$
3 × 8	$25/8 \times 71/2$
3 × 10	$25/8 \times 91/2$
3 × 12	$25/8 \times 111/2$
4 × 12	$35/8 \times 111/2$
4 × 16	$35/8 \times 151/2$
6 × 12	$51/2 \times 111/2$
6 × 16	$51/2 \times 151/2$
6 × 18	$51/2 \times 171/2$
8 × 16	$71/2 \times 151/2$
8 × 20	$71/2 \times 191/2$
8 × 24	$71/2 \times 231/2$

standard sizes after it is dressed. Quoted sizes refer to the dimensions of the wood before it is surfaced. These differences must be taken into consideration.

A good example of the dimension difference is the common 2×4. As you can see in the table, the familiar quoted size of 2×4 is the rough or nominated dimension, while the actual dressed size is $15/8 \times 35/8$ inches.

Lumber, as it comes from the sawmill, is divided into three main classes: yard lumber, structural material, and factory and shop lumber. Yard lumber is that used for ordinary construction and general building purposes. It is subdivided into the classifications of select lumber and common lumber.

Select lumber is of good appearance and finish. It is identified by the following grade names: grade A, grade B, grade C, and grade D.

Common lumber is suitable for general construction and utility purposes. It has the following grade names: No. 1 common, No. 2 common, No. 3 common, No. 4 common, and No. 5 common.

Softwood flooring is graded as select: A, B, C, and D. First-grade softwood flooring is B or better; second-grade is C or better.

Hardwood flooring is graded as Clear, Select, then No. 1 common, etc. Some types of hardwood will be graded as first-grade, second-grade, etc.

ESTIMATING BOARD FEET

The sizes of lumber and woods are standardized for ordering and handling convenience. Building and finish material sizes run 8, 10, 12, 14, 16, 18, and 20 feet in length; 2, 4, 6, 8, 10, and 12 inches in width; and 1, 2, and 4 inches in thickness. The actual width and thickness of the dressed lumber are considerably less than the standard, or quoted, width and thickness. Hardwoods, which have no standard lengths or widths, run $1/4$, $1/2$, 1, $1^1/4$, $1^1/2$, 2, $2^1/2$, 3, and 4 inches in thickness.

Plywoods are usually 4×8 feet and vary in thickness from $1/8$ to 1 inch. Stock panels are usually available in widths of 48 inches. Lengths varying by multiples of 16 inches up to 8 feet. Panel lengths run in 16-inch multiples because the accepted spacing for studs and joists is 16 inches.

The amount of lumber required for a job is measured in board feet. A board foot is a unit of measure that represents an area of 1 square foot and a thickness of 1 inch actual or nominal size. The number of board feet in a piece of lumber can be computed by the arithmetic method or by using a table.

To determine the number of board feet in one or more pieces of lumber, the following formula is used:

$$\frac{\text{pieces} \times \text{thickness} \times \text{width} \times \text{length}}{12}$$

In the equation above, the thickness and width are expressed in inches and the length in feet. As an example, here's how to find the number of board feet in a piece of lumber that is 2 inches thick, 10 inches wide, and 6 feet long:

$$\frac{1 \times 2 \times 10 \times 6}{12} = 10 \text{ board feet}$$

Rapid estimations of board feet can also be made using TABLES 2-2 or 2-3.

Table 2-2   Rapid Calculation of Board Measure

Width	Thickness	Board feet
3″	1″ or less	$1/4$ of the length
4″	1″ or less	$1/3$ of the length
6″	1″ or less	$1/2$ of the length
9″	1″ or less	$3/4$ of the length
12″	1″ or less	Same as the length
15″	1″ or less	$1^1/4$ of the length

Table 2-3   Board Feet

Nominal size (in.)	Actual length in feet								
	8	10	12	14	16	18	20	22	24
1 × 2		1²/₃	2	2¹/₃	2²/₃	3	3¹/₂	3²/₃	4
1 × 3		2¹/₂	3	3¹/₂	4	4¹/₂	5	5¹/₂	6
1 × 4	2³/₄	3¹/₃	4	4²/₃	5¹/₃	6	6²/₃	7¹/₃	8
1 × 5		4¹/₆	5	5⁵/₆	6²/₃	7¹/₂	8¹/₃	9¹/₆	10
1 × 6	4	5	6	7	8	9	10	11	12
1 × 7		5⁵/₈	7	8¹/₆	9¹/₃	10¹/₂	11²/₃	12⁵/₆	14
1 × 8	5¹/₃	6²/₃	8	9¹/₃	10²/₃	12	13¹/₃	14²/₃	16
1 × 10	6²/₃	8¹/₃	10	11²/₃	13¹/₃	15	16²/₃	18¹/₃	20
1 × 12	8	10	12	14	16	18	20	22	24
1¹/₄ × 4		4¹/₆	5	5⁵/₆	6²/₃	7¹/₂	8¹/₃	9¹/₆	10
1¹/₄ × 6		6¹/₄	7¹/₂	8³/₄	10	11¹/₄	12¹/₂	13³/₄	15
1¹/₄ × 8		8¹/₃	10	11²/₃	13¹/₃	15	16²/₃	18¹/₃	20
1¹/₄ × 10		10⁵/₁₂	12¹/₂	14⁷/₁₂	16²/₃	18³/₄	20⁵/₆	22¹¹/₁₂	25
1¹/₄ × 12		12¹/₂	15	17¹/₂	20	22¹/₂	25	27¹/₂	30
1¹/₂ × 4	4	5	6	7	8	9	10	11	12
1¹/₂ × 6	6	7¹/₂	9	10¹/₂	12	13¹/₂	15	16¹/₂	18
1¹/₂ × 8	8	10	12	14	16	18	20	22	24
1¹/₂ × 10	10	12¹/₂	15	17¹/₂	20	22¹/₂	25	27¹/₂	30
1¹/₂ × 12	12	15	18	21	24	27	30	33	36
2 × 4	5¹/₃	6²/₃	8	9¹/₃	10¹/₃	12	13¹/₃	14²/₃	16
2 × 6	8	10	12	14	16	18	20	22	24
2 × 8	10²/₃	13¹/₃	16	18²/₃	21¹/₃	24	26²/₃	29¹/₃	32
2 × 10	13¹/₃	16²/₃	20	23¹/₃	26²/₃	30	33¹/₃	36²/₃	40
2 × 12	16	20	24	28	32	36	40	44	48
3 × 6	12	15	18	21	24	27	30	33	36
3 × 8	16	20	24	28	32	36	40	44	48
3 × 10	20	25	30	35	40	45	50	55	60
3 × 12	24	30	36	42	48	54	60	66	72
4 × 4	10²/₃	13¹/₃	16	18²/₃	21¹/₃	24	26²/₃	29¹/₃	32
4 × 6	16	20	24	28	32	36	40	44	48
4 × 8	21¹/₃	26²/₃	32	37¹/₃	42²/₃	48	53¹/₃	58²/₃	64
4 × 10	26²/₃	33¹/₃	40	46²/₃	53¹/₃	60	66²/₃	73¹/₃	80
4 × 12	32	40	48	56	64	72	80	88	96

BUYING HARDWOOD FLOORING

As with most consumer products, how you purchase can be as important to the job as what you purchase. If you buy inferior hardwood flooring and don't select a retailer who will stand behind what is sold, you can't expect a superior-quality floor.

An important and often overlooked fact is that when you buy any product, you also are buying whatever support service the retailer offers.

At the big chain store that has some hardwood flooring on sale, you will probably know more about the product you buy than will the clerk who sells it to you. Unless you are an expert in the product you are selecting, avoid such retailers. You will be farther ahead if you buy from a retailer who can support his merchandise with product knowledge and assistance.

How can you purchase hardwood flooring from specialized retailers and still get a good price? First, select a retailer from whom you would feel comfortable buying if you had more money. Then let them know that you don't have to buy it immediately and that you would prefer to wait until you could earn a savings of 25% or more. Most retailers understand and will either work out some type of discount nearing or surpassing this figure, or will suggest when that price discount will be available.

To make the best purchase, you must be sure that there are no hidden costs and know exactly what is included with your purchase. Once you know which type of hardwood flooring to select, how it will be installed, and approximately how much you will need, make a list of the components you will need to buy—flooring, fasteners, adhesives, tools, finishes, and special equipment rentals. Then you can make sure that the total price really is the total.

You also need to know what you're buying. You should know the characteristics of wood and understand the different types of joints and flooring patterns. This information is offered in Chapter 1. Later in this chapter, you will find out about tools and fasteners. Read about them before you go on your shopping trip.

One method that is used by smart consumers to ensure that what they bought is what they thought they bought is to defer partial payment. Make arrangements to pay part of the bill when the order comes in, part on delivery after an inspection and inventory, and the final amount 30 days after delivery. The flooring retailer who often works with local contractors is accustomed to such terms. These terms offer you the opportunity to verify the quantity and quality of the hardwood flooring and related materials with recourse.

Remember that a bargain is only a bargain when the job is complete. What may have initially cost less to purchase may eventually cost more. One of the reasons you are doing it yourself is to save money. By purchasing your hardwood flooring as a smart consumer, you will save money and have a better quality floor when you're done.

SELECTING AND USING HARDWOOD FLOORING TOOLS

There are a variety of hand and power tools available to the do-it-yourselfer who wants to install hardwood flooring inexpensively and efficiently. Figure 2-2 illustrates the basic tools that are suggested by one hardwood-flooring manufacturer. They include a handsaw or power saw, hammer, crowbar, square, chalk line, wedge, adhesive, and a special installation tool.

One of the most important tools for a hardwood-floor installation is the measuring tool. The ability to accurately lay out and measure a room

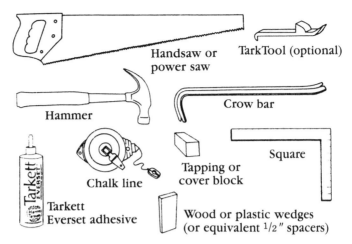

2-2 Typical tools needed for hardwood flooring installation (courtesy Harris-Tarkett, Inc.).

depends on the correct use of the measuring tools and the ease with which the graduations on the tools can be read. While each measuring tool is used for a specific purpose, they are all graduated according to the same system of linear measure.

Figure 2-3 shows the types of rules and tapes that are commonly used by builders and do-it-yourselfers. Of all the measuring tools, one of the most practical is the steel rule. This rule is usually 6 or 12 inches in length. Steel rules may be flexible or inflexible, but the thinner the rule, the easier it is to measure accurately because the division marks are closer to the work.

Generally, a rule has four sets of graduations—one on each edge of each side. The longest lines represent the inch marks. On one edge, each space represents 1/8 inch. The other edge of this side is divided by 1/16-inch marks. The 1/4-inch and 1/2-inch marks are commonly made longer than the smaller division marks to facilitate counting. The graduations are not, as a rule, numbered individually, however, since they are sufficiently far apart to be counted without difficulty. The opposite side is

2-3 Common types of rules and tapes.

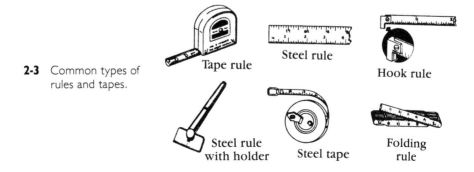

similarly divided into 32 and 64 spaces per inch. It is common practice to number every fourth division of these graduations for easier reading.

There are many variations of the common rule. Sometimes the graduations are on one side only. Sometimes the graduations are added across one end for measuring in narrow spaces. And sometimes only the first inch is divided into 64ths, with the remaining inches divided into 32nds or 16ths.

To measure lengths that are greater than 18 inches, folding steel, wood, or aluminum rules can be used. They are usually 2 to 6 feet long. The folding rules cannot be relied on for extremely accurate measurements because a certain amount of play develops at the joints after they have been used for awhile.

Steel tapes are made from 6 to about 100 feet in length. The shorter lengths are frequently made with a curved cross section so that they are flexible enough to roll up, but remain rigid when extended. Long, flat tapes require support over their full length when they are used to measure, or the natural sag will cause an error in the reading.

The flexible-rigid tapes are usually contained in metal cases. They may wind themselves back into the case when a button is pressed or they may be easily wound with a crank. A hook is provided at one end so that the tape may be hooked over the object being measured. This allows one person to handle the tape without assistance. On some models, the outside of the case can be used as one end of the tape when inside dimensions are being measured.

Rules and tapes should be handled carefully and kept lightly oiled to prevent rust. Never allow the edges of measuring devices to become nicked from being striked by hard objects. They should preferably be kept in a wooden box when not in use.

To avoid kinking the tapes, pull them straight out from their cases. Do not bend them backward. With the windup type, always turn the crank clockwise. Turning it backward will kink or break the tape. With the spring-wind type, guide the tape by hand. If it is allowed to snap back, it may be kinked, twisted, or otherwise damaged.

USING A TAPE RULE

Notice in FIG. 2-4 that the circle near the hook at the end of the perpendicular rule is attached so that it is free to move slightly. When an outside dimension is taken by hooking the end of the rule over the edge, the hook will locate the end of the rule even with the surface from which the measurement is being taken.

To measure an inside dimension with a tape rule, extend the rule between the surfaces as shown, then take a reading at the point on the scale where the rule enters the case and add 2 inches. The 2 inches are the width of the case. The total is the inside dimension.

To measure an outside dimension with a tape rule, hook the rule over the edge of the stock. Pull the tape out until it projects far enough from the case to permit you to measure the required distance. The hook is

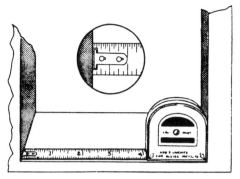

2-4 Measuring inside dimensions with a tape rule.

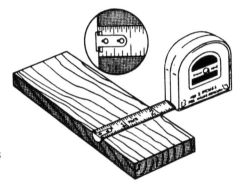

2-5 Measuring outside dimensions with a tape rule.

designed so that it will locate the end of the rule at the surface from which the measurement is being taken (FIG. 2-5). When you are taking a measurement of length, hold the tape parallel to the lengthwise edge.

For measuring widths, the tape should be at right angles to the lengthwise edge. Read the dimension of the rule exactly at the edge of the piece that is to be measured. In this case, it may be necessary to butt the end of the tape against another surface or to hold the rule at a starting point from which a measurement is to be taken.

USING A FRAMING SQUARE

Of all the layout tools in the woodworker's kit, the framing square is the most generally useful.

When you lay out 90- and 45-degree angles with a framing square, the lumber you work with should be squared on the ends. This will make it necessary to lay out a line at a 90-degree angle with respect to the edge of the board. This line should be as close to the end of the board as possible to avoid undue material waste. When doing this job with a framing square, place the blade of the square along one edge of the board and mark along the outside edge of the tongue, as shown in FIG. 2-6.

2-6   Using a framing square.

USING A COMBINATION SQUARE

Another versatile tool for measuring, marking, and preparing hardwood floor strips or blocks for cutting is the combination square. To square a line on stock with a combination square, place the squaring head on the edge of the stock, as shown in FIG. 2-7, and draw the line along either edge of the blade. The line will be square with the edge of the stock against which the squaring head is held. The angle between the line and the edge will be 90 degrees.

To lay out a 45-degree angle on stock with a combination square, place the squaring head on the edge of the stock, as shown in FIG. 2-8, and draw the line along either edge of the blade. The line will be at a 45-degree angle to the edge of the stock against which the squaring head is held.

To test the trueness of 90-degree angles with a combination square, hold the body of the square in contact with one surface of the 90-degree angle and bring the blade into contact with the other (FIG. 2-9). When making this test, have the square between yourself and a good source of light. If the angle is a true 90-degree angle, no light will be visible between the blade and the surface of the work.

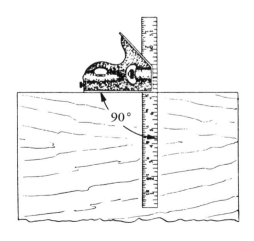

90°

2-7   Squaring stock with a combination square.

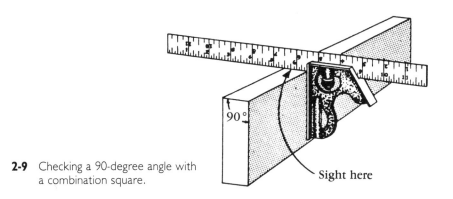

2-8 Making 45-degree cuts with a combination square.

2-9 Checking a 90-degree angle with a combination square.

The chalk line is one other useful tool that could be mentioned here (FIG. 2-10). A chalk line is a white, twisted mason's line that consists of a reel, line, and chalk. The line is coated with chalk. When it is stretched taut between the points that are to be connected by a straight line and held just off the surface, the chalk line will make a straight guideline when it is snapped. The chalk line can be very useful when making diagonal cuts across hardwood floor boards.

HANDSAWS

Anyone who works with wood uses a large variety of hand tools. Of these, woodcutting hand tools are often the most used—and abused.

2-10 Using a chalk line.

2-11 Using a common carpenter's saw.

The most common carpenter's handsaw (FIG. 2-11) consists of a steel blade with a handle at one end. The blade is narrower at the end opposite the handle, which is called the point or toe. The end of the blade nearest the handle is called the heel. One edge of the blade has teeth that act as two rows of cutters. When the saw is used, these teeth cut two parallel grooves close together. The chips or sawdust, are pushed out from between the grooves, or kerfs, by the beveled part of the teeth. The teeth are bent alternately to one side or the other to make the kerf wider than the thickness of the blade. This bending is called the set of the teeth.

The number of teeth per inch, the size and shape of the teeth, and the amount of set determine how the saw should be used and the type of material that it should cut. Carpenter's handsaws are described by the number of points per inch. There is always one more point than there are teeth per inch. A number stamped near the handle gives the number of points of the saw.

Woodworking handsaws consist of ripsaws and crosscut saws that are designed for general cutting. Ripsaws are used for cutting with the grain, and crosscut saws are used to cut across the grain. The major difference between a ripsaw and a crosscut saw is the shape of the teeth. A ripsaw tooth has a square-faced, chisel cutting edge (FIG. 2-12). It does a good job of cutting with the grain, which is called ripping, but a poor job of cutting across the grain, which is called crosscutting. A crosscut saw tooth has a beveled, knife-like cutting edge, like the one in FIG. 2-13. It does a good job of cutting across the grain, but a poor job of cutting with the grain. Figure 2-14 illustrates a combination saw blade that offers the qualities of both the ripsaw and the crosscut saw.

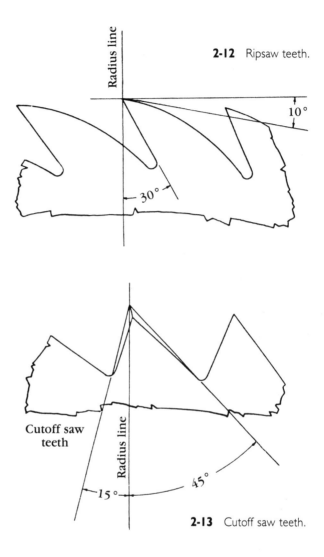

2-12 Ripsaw teeth.

10°

30°

Radius line

2-13 Cutoff saw teeth.

Cutoff saw teeth

Radius line

15°

45°

2-14 Combination saw teeth.

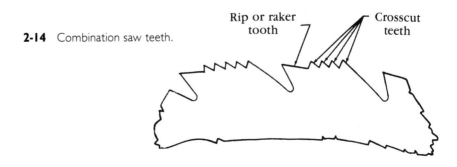

Rip or raker tooth

Crosscut teeth

Special handsaws

The more common types of saws, which are used for special purposes, are shown in FIG. 2-15. They can be useful as you notch hardwood floor strips or blocks around pipes or make critical cuts.

The backsaw is a crosscut saw that is designed for sawing a perfectly straight line across the face of a piece of stock. A heavy steel backing along the top of the blade keeps the blade perfectly straight. The dovetail saw is a special type of backsaw that has a thin, narrow blade and a chisel-like handle.

The compass saw is a long, narrow, tapering ripsaw that was designed to cut out circular or other nonrectangular sections from within the margin of a board or panel. A hole is bored near the cutting line to start the saw. The keyhole saw is simply a finer, narrower compass saw. The coping saw is used to cut along the curved lines, as shown in FIG. 2-15.

Caring for handsaws

Some of the right and wrong methods of using and caring for a handsaw are shown in FIG. 2-16. A saw that is not being used should be hung up or stored in a toolbox. A toolbox that is designed for holding saws has notches that hold them on edge with the teeth facing up. If you store saws loose in a toolbox, the saw teeth may become dulled or bent from contact with the other tools.

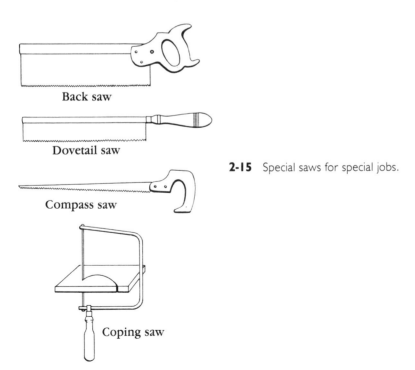

Back saw

Dovetail saw

Compass saw

Coping saw

2-15   Special saws for special jobs.

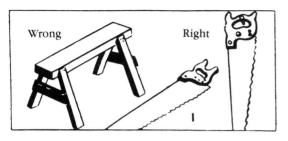

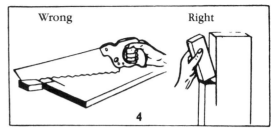

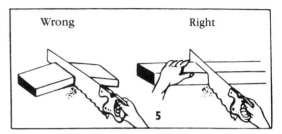

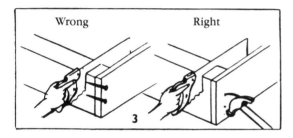

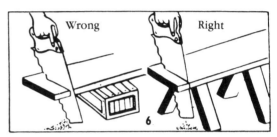

1 When work is complete, hang up the saw.

2 Do not pile tools on top of the bench so as to distort blade.

3 Look carefully over repair or alteration work; see that all nails are removed to avoid cutting into metal.

4 Strips of waste should not be twisted off with blade, but broken off with hand or mallet.

5 Supporting the waste side of work will prevent splitting off.

6 Raise the work to a height sufficient to keep the blade from striking the floor. If the work cannot be raised, limit the stroke.

2-16 Caring for handsaws.

Before you use a saw, be sure that there are no nails or other edge-destroying objects in the line of the cut. When sawing out a strip of waste, don't break out the strip by twisting the saw blade. Doing this dulls the saw and may spring or break the blade.

Be sure that the saw will go through the full stroke without striking the floor or some other object. If the work cannot be raised high enough to obtain full clearance for the saw, you must carefully limit the length of each stroke.

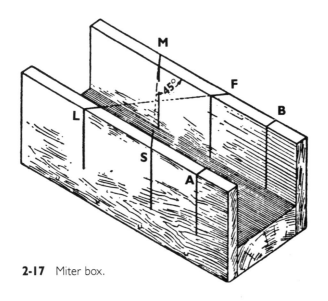

2-17 Miter box.

Sawing aids

A miter box (FIG. 2-17) permits you to saw a piece of stock at a given angle without laying out a line. The figure shows a common type of wooden 45-degree miter box. Stock can be cut at 45 degrees by placing the saw in cuts M-S and L-F, or at 90 degrees by placing the saw in cuts A-B.

The sawhorse (FIG. 2-18) might be called the carpenter's portable workbench and scaffold. If you don't already have a good sawhorse, follow the directions in the illustration and make one. Not only is the sawhorse practical, but making one is a good exercise in how to use your handsaw and measuring tools. It also will help you gain an understanding of wood.

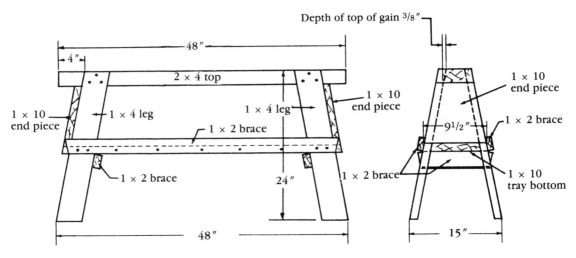

2-18 Dimensions for making your own sawhorse.

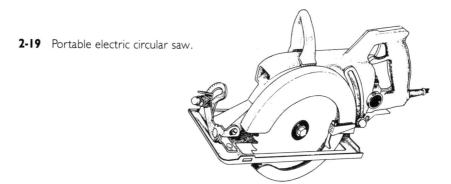

2-19 Portable electric circular saw.

PORTABLE CIRCULAR SAW

The portable, electric circular saw (FIG. 2-19) is one of the most popular tools for the do-it-yourselfer. It can be used for any purpose that would take a handsaw, while offering the advantage of speed. The size of a portable circular saw is designated by the maximum diameter of the blade in inches that it will support within its guard.

To make an accurate rip cut, the ripping guide (FIG. 2-20) is set a distance away from the saw that is equal to the width of the strip that is to be ripped off and placed against the edge of the piece as a guide for the saw. When the cut is finished, the ripping guide is turned upside down so that it will be out of the way.

Circular saws use circular blades. The types of blades are the same as those available on handsaws: ripping, crosscutting, compass, and combination.

Power saw safety

All portable, power-driven saws should be equipped with guards that automatically adjust themselves so that none of the teeth protrude above the work. The guard over the blade should be adjusted so that it slides out of its recess and covers the blade to the depth of the teeth when the saw is lifted from the work. Goggles or face shields should be worn while using the saw and while cleaning up the debris afterwards.

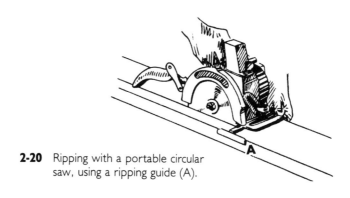

2-20 Ripping with a portable circular saw, using a ripping guide (A).

Saws should be grasped with both hands and held firmly against the work. Care should be taken that the saw does not break away, which could cause an injury. The blade should be inspected at frequent intervals and always after it has locked, pinched, or burned. Pull the plug to break the electrical connection before this examination. Do not overload the saw motor by pushing too hard or cutting stock that is too heavy for the saw.

Before using the saw, carefully examine the material that is to be cut. It should be free of nails or other metal substances. Cutting into or through knots should be avoided as much as possible.

The electric plug should be pulled before any adjustments or repairs are made to the saw, including blade changes.

CARPENTER'S LEVEL

The carpenter's level (FIG. 2-21) determines the levelness of a surface and allows you to sight level lines. It may be used directly on the surface or with a straightedge (FIG. 2-22). The levelness of an object is determined by the bubbles that are suspended within glass tubes that are parallel to one or more surfaces of the level (FIG. 2-23).

To level a surface, such as the workbench in FIG. 2-24, set the carpenter's level on the benchtop parallel to the front edge of the bench. Notice that the level may have as many as three or more pairs of glass vials. Regardless of the position of the level, always watch the bubble in the bottom vial of the horizontal pair. Shim or wedge up the end of the bench

2-21   Carpenter's level.

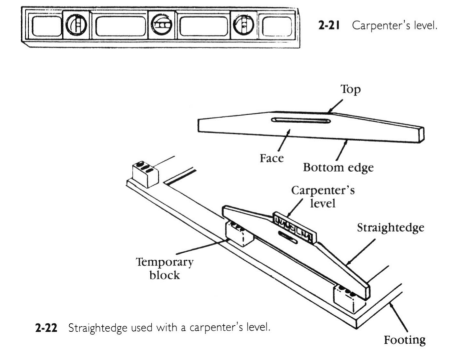

Top

Face

Bottom edge

Carpenter's level

Straightedge

Temporary block

2-22   Straightedge used with a carpenter's level.

Footing

Horizontal bubble tube

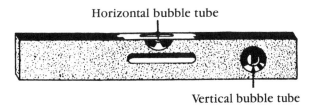

Vertical bubble tube

2-23 Smaller carpenter's level with horizontal and vertical bubble tubes.

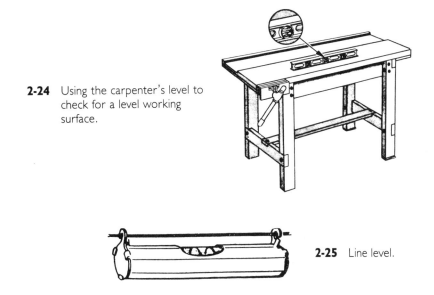

2-24 Using the carpenter's level to check for a level working surface.

2-25 Line level.

that will return the bubble to the center of its vial. Recheck the first position of the level before you secure the shims or wedges. These principles can easily be applied to the leveling of flooring material with a carpenter's level and shims.

The line level (FIG. 2-25) has a spirit bubble that shows levelness as it is hung from a line. Placement halfway between the points that are to be leveled gives the greatest accuracy.

CLAW HAMMER

The carpenter's curved-claw nail hammer (FIG. 2-26) is a steel-headed, wooden-handled tool that is used to drive nails, wedges, and dowels. The claw, which is at one end of the head, is a two-pronged arch that is used to pull nails out of the piece of wood. The other parts of the head are the eye and face.

The face may be flat, in which case it is called a plain face (FIG. 2-27). The beginning woodworker will find the plain-faced hammer the easiest hammer to use to learn to drive nails. With this hammer, however, it is dif-

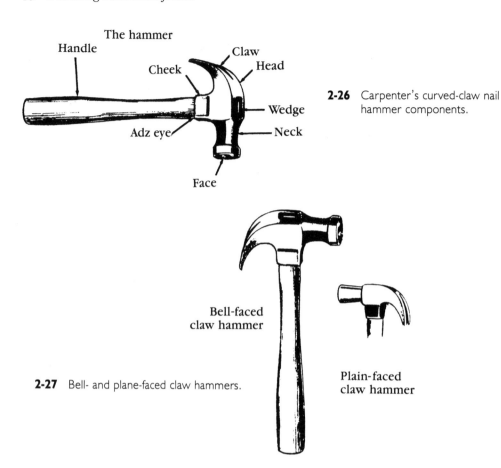

The hammer

Handle Cheek Claw Head Wedge Neck Adz eye Face

2-26 Carpenter's curved-claw nail hammer components.

Bell-faced claw hammer

Plain-faced claw hammer

2-27 Bell- and plane-faced claw hammers.

ficult to drive the nailhead flush with the surface of the work without leaving hammer marks on the surface.

The face of a hammer may also be slightly rounded or convex, in which case it is called bell-faced. The bell-faced hammer is generally used in rough work. When handled by an expert, it can drive the nailhead flush with the surface of the work without damaging the surface.

To use a hammer, grasp the handle so that the end is flush with the lower edge of your palm (FIG. 2-28). Keep your wrist limber and relaxed.

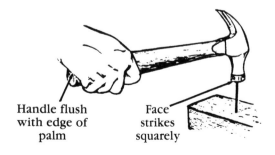

Handle flush with edge of palm

Face strikes squarely

2-28 Correct way to use a hammer.

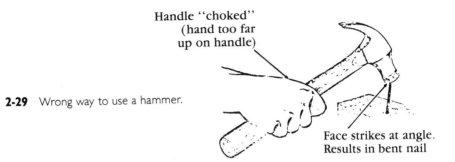

Handle "choked"
(hand too far
up on handle)

2-29 Wrong way to use a hammer.

Face strikes at angle.
Results in bent nail

Grasp the nail with the thumb and forefinger of your other hand and place the point at the exact spot where it is to be driven. Unless the nail is to be purposely driven at an angle, it should be perpendicular to the surface of the work. Strike the nailhead squarely and keep your hand level with the head of the nail. To drive, first rest the face of the hammer on the head of the nail. Then raise the hammer slightly and give the nail a few light taps to start it and fix the aim. Then take your fingers away from the nail and drive the nail with firm blows with the center of the hammer face. The wrong way to drive a nail is shown in FIG. 2-29.

SELECTING AND USING HARDWOOD FLOOR FASTENERS

Nailing two pieces of wood together is one of the most common tasks in carpentry and one that you'll be able to practice many times as you install your hardwood floors. If the joint doesn't hold or the wood splits, it is generally because the installer didn't observe a few basic rules for nailing.

First of all, if the joint is to hold properly, the nail must be long enough. A good rule to follow here is to select a nail three times the length of the thickness of the wood that is to be nailed. If the nail is too short, it cannot hold properly. If it is too long, the increased diameter may split the wood. There will be more information on how to select nails later in this chapter.

A few properly spaced nails will hold better than many nails that are put in at one point. Improper spacing or too many nails will split the

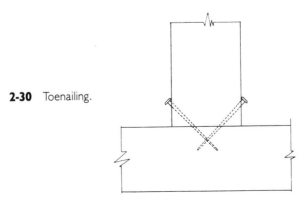

2-30 Toenailing.

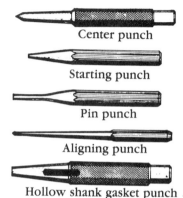

Putty

2-31   Filling over a set finish nail.

Center punch

Starting punch

2-32   Commonly used punches.

Pin punch

Aligning punch

Hollow shank gasket punch

wood and not add strength to the joint. A nail that is driven in at an angle, which is called toenailing (FIG. 2-30), will provide a stronger joint than a nail that is driven straight down.

When the end of the nail extends through the second piece of wood, it should be clinched or bent over. Although clinching the nail with the grain will give a smoother surface, clinching across the grain will give more strength.

The head of a finishing nail should be set below the surface of the wood (FIG. 2-31) with a nail set and the resulting hole filled with putty or plastic wood. Whenever a finished appearance is desired, drive the nail almost to the surface with the hammer and finish the job with a nail set (FIG. 2-32). This method will keep you from striking the wood with the face of the hammer and denting it.

When you use a claw hammer to pull out nails, insert a block of wood under the hammer to provide more leverage and to prevent the hammer from damaging the wood. If the nail has been clinched, it should be straightened before any attempt is made to remove it.

Considering how inexpensive nails are, it is a waste of time and materials to try to straighten nails for reuse. Once a nail has been used and bent back and forth, it has lost much of its holding power. Moreover, it is almost impossible to straighten a nail perfectly. The result is a bent nail that must be removed, possibly resulting in damage to the wood.

NAILS

Figure 2-33 shows the more common types of wire nails. The brad and the finish nail both have a deep countersink head that is designed to be set

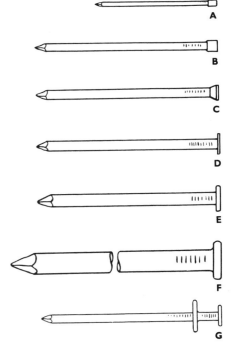

2-33 Common types of wire nails: (A) brad, (B) finish nail, (C) casing nail, (D) box nail, (E) common nail, (F) spike (larger than 60d), (G) duplex head nail.

below the surface of the wood. The casing nail has a flat countersink head, which may be driven flush and left that way or which may also be set. The other nails shown are all flat head nails.

The common nail is the one most widely used in general wood construction. Nails with large flat heads are used for nailing roofing paper, plaster board, and similar thin or soft materials. Duplex or double-headed nails are used for nailing temporary structures, such as scaffolds, that will eventually be dismantled. A duplex nail has an upper and lower head. The nail is driven to the lower head so it can be easily withdrawn by setting the claw of a hammer under the upper head.

Besides nails with the usual type of shank, which is round, there are various special-purpose nails with shanks of other shapes. Nails with square, triangular, longitudinally grooved, and spirally grooved shanks have a much greater holding power than wire nails of the same size. One you may come across is the flooring nail (FIG. 2-34), which is driven diagonally through the tongue of the board and into the subfloor. The availability of these nails, however, is somewhat limited for the do-it-yourselfer and the skill required for placement is high, so they are usually not used.

The lengths of the most commonly used nails are designated by the penny system. This system originated in England where the abbreviation for the word penny is the letter d. Thus, the expression two-penny nail is written 2d nail. The thickness of a nail increases with the penny size and the number of nails per pound decreases. A box of casing nails of the same common nail penny size is thinner. Consequently, it takes more boxes to get the same weight.

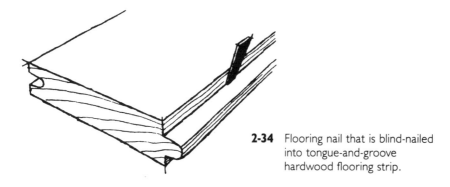

2-34 Flooring nail that is blind-nailed into tongue-and-groove hardwood flooring strip.

The penny sizes and corresponding lengths, thicknesses (in gauge sizes), and numbers per pound of the most commonly used nails are shown in TABLE 2-4. Relative sizing is shown in FIG. 2-35. Recommended nailing methods and sizes for specific home construction projects are given in TABLE 2-5.

The lengths of nails that are larger than 60d, which are called spikes, are designated in inches. Nails smaller than 2d are designated in fractions of an inch.

WOOD SCREWS

Wood screws (FIG. 2-36) are designated by the type of head and the material, such as flathead brass or roundhead steel. Most wood screws are made of either steel or brass, but there are copper and bronze wood screws as well. To distinguish the ordinary type of head from the Phillips head, the former is called a slotted head. A lag screw is a heavy iron screw that has a square bolt-type head. Lag screws are used mainly for fastening heavy timbers, but they can also be used for the installation of hardwood plank floors.

The size of an ordinary wood screw is designated by the length and the body diameter, or unthreaded part, of the screw (FIG. 2-37). Body diameters are designated by gauge numbers that run from 0, for about a $1/16$-inch diameter, to 24, for about a $3/8$-inch diameter. Lengths range from $1/4$ inch to 5 inches. The length and gauge numbers are printed on the box, as $1 1/4$-9. This means a 9 gauge screw $1 1/4$ inches long.

Note that for a nail a large gauge number means a small nail, but for a screw, a large gauge number means a large screw. TABLES 2-6 and 2-7 will guide you through the sizes of wood screws.

Figure 2-38 illustrates how screws can be countersunk through hardwood flooring and the subflooring. First, drill the body hole completely through the flooring. Then drill the starter hole, which is a little less than the diameter of the wood screw. Finally, if a flathead wood screw or ovalhead wood screw is to be used, countersink the body hole as shown.

Thus far in this chapter, you've learned a great deal about planning hardwood floors and the tools you'll need to install a hardwood floor.

Table 2-4 Sizes of Commonly Used Nails

Common wire nails

Size	Length	Gauge	Approx. No. to Lb.	Size	Length	Gauge	Approx. No. to Lb.
2d	1 In.	No. 15	876	10d	3 In.	No. 9	69
3d	1 1/4	14	568	12d	3 1/4	9	63
4d	1 1/2	12 1/2	316	16d	3 1/2	8	49
5d	1 3/4	12 1/2	271	20d	4	6	31
6d	2	11 1/2	181	30d	4 1/2	5	24
7d	2 1/4	11 1/2	161	40d	5	4	18
8d	2 1/2	10 1/4	106	50d	5 1/2	3	14
9d	2 3/4	10 1/4	96	60d	6	2	11

Flooring brads

Size	Length	Gauge	Approx. No. to Lb.
6d	2 In.	No. 11	157
7d	2 1/4	11	139
8d	2 1/2	10	99
9d	2 3/4	10	90
10d	3	9	69
12d	3 1/4	8	54
16d	3 1/2	7	43
20d	4	6	31

Finishing nails

Size	Length	Gauge	Approx. No. to Lb.
2d	1 In.	No. 16 1/2	1351
3d	1 1/4	15 1/2	807
4d	1 1/2	15	584
5d	1 3/4	15	500
6d	2	13	309
7d	2 1/4	13	238
8d	2 1/2	12 1/2	189
9d	2 3/4	12 1/2	172
10d	3	11 1/2	121
12d	3 1/4	11 1/2	113
16d	3 1/2	11	90
20d	4	10	62

Smooth & barbed box nails

Size	Length	Gauge	Approx. No. to Lb.
2d	1 In.	No. 15 1/2	1010
3d	1 1/4	14 1/2	635
4d	1 1/2	14	473
5d	1 3/4	14	406
6d	2	12 1/2	236
7d	2 1/4	12 1/2	210
8d	2 1/2	11 1/2	145
9d	2 3/4	11 1/2	132
10d	3	10 1/2	94
12d	3 1/4	10 1/2	88
16d	3 1/2	10	71
20d	4	9	52
30d	4 1/2	9	46
40d	5	8	35

Casing nails

Size	Length	Gauge	Approx. No. to Lb.
2d	1 In.	No. 15 1/2	1010
3d	1 1/4	14 1/2	635
4d	1 1/2	14	473
5d	1 3/4	14	406
6d	2	12 1/2	236
7d	2 1/4	12 1/2	210
8d	2 1/2	11 1/2	145
9d	2 3/4	11 1/2	132
10d	3	10 1/2	94
12d	3 1/4	10 1/2	87
16d	3 1/2	10	71
20d	4	9	52
30d	4 1/2	9	46

Table 2-5   Recommended Nailing for Houses

Joining	Nailing method	Number	Size	Placement
Header to joist	End-nail	3	16d	
Joist to sill or girder	Toenail	2–3	10d or 8d	
Header and stringer joist to sill	Toenail		10d	16 inches on center.
Bridging to joist	Toenail each end	2	8d	
Ledger strip to beam, 2 inches thick		3	16d	At each joist.
Subfloor, boards:				
1 by 6 inches and smaller		2	8d	To each joist.
1 by 8 inches		3	8d	To each joist.
Subfloor, plywood:				
At edges			8d	6 inches on center.
At intermediate joists			8d	8 inches on center.
Subfloor (2 by 6 inches, T&G) to joist or girder	*Blind-nail* (casing) and face-nail.	2	16d	
Soleplate to stud, horizontal assembly	End-nail	2	16d	At each stud.
Top plate to stud	End-nail	2	16d	
Stud to soleplate	Toenail	4	8d	
Soleplate to joist or blocking	*Face-nail*		16d	16 inches on center.
Doubled studs	Face-nail, stagger		10d	16 inches on center.
End stud of intersecting wall to exterior wall stud	Face-nail		16d	16 inches on center.
Upper top plate to lower top plate	Face-nail		16d	16 inches on center.
Upper top plate, laps and intersections	Face-nail	2	16d	
Continous header, 2 pieces, each edge			12d	12 inches on center.
Ceiling joist to top wall plates	Toenail	3	8d	
Ceiling joist laps at partition	Face-nail	4	16d	
Rafter to top plate	Toenail	2	8d	
Rafter to ceiling joist	Face-nail	5	10d	
Rafter to valley or hip rafter	Toenail	3	10d	
Ridge board to rafter	End-nail	3	10d	
Rafter to rafter through ridge board	{Toenail	4	8d	
	{Edge-nail	1	10d	
Collar beam to rafter:				
2-inch member	Face-nail	2	12d	
1-inch member	Face-nail	3	8d	
1-inch diagonal let-in brace to each stud and plate (4 nails at top).		2	8d	
Built-up corner studs:				
Studs to blocking	Face-nail	2	10d	Each side.
Intersecting stud to corner studs	Face-nail		16d	12 inches on center.
Built-up girders and beams, 3 or more members	Face-nail		20d	32 inches on center, each side.
Wall sheathing:				
1 by 8 inches or less, horizontal	Face-nail	2	8d	At each stud.
1 by 6 inches or greater, diagonal	Face-nail	3	8d	At each stud.
Wall sheathing, vertically applied plywood:				
⅜ inch and less thick	Face-nail		6d	}6-inch edge.
½ inch and over thick	Face-nail		8d	}12-inch intermediate.
Wall sheathing, vertically applied fiberboard:				
½ inch thick	Face-nail			1½-inch roofing nail.[1]
25/32 inch thick	Face-nail			1¾-inch roofing nail.[1]
Roof sheathing, boards, 4-, 6-, 8-inch width	Face-nail	2	8d	At each rafter.
Roof sheathing plywood:				
⅜ inch and less thick	Face-nail		6d	}6-inch edge and 12-
½ inch and over thick	Face-nail		8d	} inch intermediate.

[1] 3-inch edge and 6-inch intermediate.

Now you will see how a home is constructed from the ground up, with emphasis on how it is prepared for the installation of hardwood flooring.

MASTICS AND TROWELS

There are several types of mastic available that are satisfactory for use when laying hardwood floors. Hot asphalt is generally used only for lay-

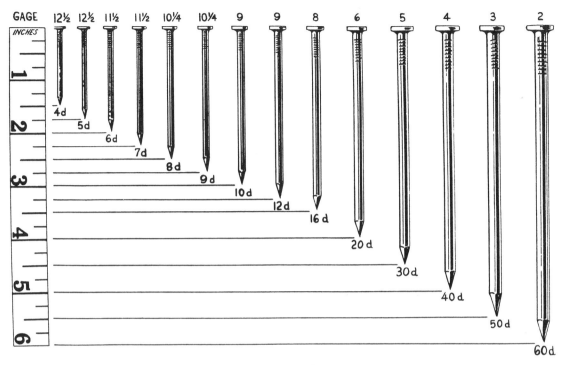

2-35 Relative nail sizes.

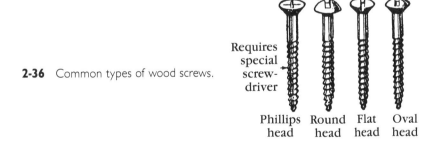

2-36 Common types of wood screws.

Requires
special
screw-
driver

Phillips   Round   Flat   Oval
head   head   head   head

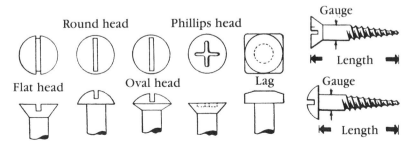

Round head   Phillips head   Gauge

Flat head   Oval head   Lag   Length

Gauge

Length

2-37 Wood screw heads and components.

Table 2-6 Screw Threads per Inch

Diameter			Threads Per Inch		
No.	Inch	Decimal Equivalent	NC	NF	EF
0	- - - -	.0600	- - -	80	- - -
1	- - - -	.0730	64	72	- - -
2	- - - -	.0860	56	64	- - -
3	- - - -	.0990	48	56	- - -
4	- - - -	.1120	40	48	- - -
5	- - - -	.1250	40	44	- - -
6	- - - -	.1380	32	40	- - -
8	- - - -	.1640	32	36	- - -
10	- - - -	.1900	24	32	40
12	- - - -	.2160	24	28	- - -
- - -	1/4	.2500	20	28	36
- - -	5/16	.3125	18	24	32
- - -	3/8	.3750	16	24	32
- - -	7/16	.4375	14	20	28
- - -	1/2	.5000	13	20	28
- - -	9/16	.5625	12	18	24
- - -	5/8	.6250	11	18	24
- - -	3/4	.7500	10	16	20
- - -	7/8	.8750	9	14	20
- - -	1	1.0000	8	14	20

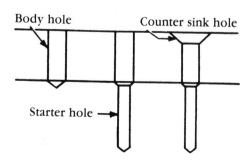

Body hole   Counter sink hole

Starter hole →

2-38 Countersinking holes for screws when attaching hardwood flooring.

Table 2-7 Screw Sizes and Dimensions

Length (in.)	Size numbers																					
	0	1	2	3	4	5	6	7	8	9	10	11	12	13	14	15	16	17	18	20	22	24
1/4	x	x	x	x																		
3/8	x	x	x	x	x	x	x	x	x	x												
1/2		x	x	x	x	x	x	x	x	x	x	x	x									
5/8		x	x	x	x	x	x	x	x	x	x	x	x		x							
3/4			x	x	x	x	x	x	x	x	x	x	x		x		x					
7/8			x	x	x	x	x	x	x	x	x	x	x		x		x					
1				x	x	x	x	x	x	x	x	x	x		x		x		x	x		
1 1/4					x	x	x	x	x	x	x	x	x		x		x		x	x		x
1 1/2					x	x	x	x	x	x	x	x	x		x		x		x	x		x
1 3/4						x	x	x	x	x	x	x	x		x		x		x	x		x
2						x	x	x	x	x	x	x	x		x		x		x	x		x
2 1/4						x	x	x	x	x	x	x	x		x		x		x	x		x
2 1/2						x	x	x	x	x	x	x	x		x		x		x	x		x
2 3/4							x	x	x	x	x	x	x		x		x		x	x		x
3							x	x	x	x	x	x	x		x		x		x	x		x
3 1/2									x	x	x	x	x		x		x		x	x		x
4									x	x	x	x	x		x		x		x	x		x
4 1/2													x		x		x		x	x		x
5														x	x		x		x	x		x
6															x		x		x	x		x
Threads per inch	32	28	26	24	22	20	18	16	15	14	13	12	11		10		9		8	8		7
Diameter of screw (in.)	.060	.073	.086	.099	.112	.125	.138	.151	.164	.177	.190	.203	.216		.242		.268		.294	.320		.372

ing screeds on concrete, and then the screeds must be positioned immediately after the mastic is poured. Cutback asphalt, chlorinated solvent, and petroleum-based solvent mastics are all applied cold and are always used for laying block and parquet floors. Cutback asphalt is also used to hold a membrane damp-proofing. Follow the manufacturers' instructions on coverage, drying time, and ventilation.

Trowels usually have both straight and notched edges. The notched edge is for use where a correct mastic thickness is specified. Both mastic and trowels are available from flooring manufacturers and distributors.

HOME CONSTRUCTION BASICS

One of the first essentials in house construction is to select the most desirable property site for its location. A lot in a smaller city or community presents few problems. The front setback of the house and side yard distances are either controlled by local regulations or governed by the other houses in the neighborhood.

Foundation

The footing is located at the base of the foundation (FIG. 2-39). It is made of concrete and is wider than the actual foundation so that the weight of

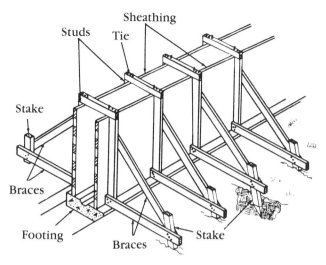

2-39 Foundation and footing.

the house will be distributed over a greater area. If the footing is not the right size for the weight of the house and the soil conditions, it will sink and the house will settle.

The foundation is the masonry which sits on top of the footing and supports the weight of the house. It also provides the walls for the basement. The foundation can be made of stone, cement, cinder blocks, poured concrete, or any other material that can sustain a considerable load.

2-40 Foundation prior to sill installation.

2-41 Close up of sill bolts embedded in foundation.

2-42 Sill plate installed with sill bolts.

The sills (FIGS. 2-40 through 2-42) are the wood or steel beams that are attached to the top of the foundation. The house is built up from the beams.

Girders (FIGS. 2-43 through 2-45) are the large beams that run between opposite sills. They are used to provide additional support for the frame of the house, as well as to carry the flooring.

2-43 Flooring girders are installed on foundation piers.

2-44 Girders fit into notches in the perimeter foundation.

The floor joists (FIGS. 2-46 and 2-47) are the beams that run across the sills and provide a base for the flooring. Floor joists are generally made of 2×10 or 2×8-inch lumber, depending on the distance they must span. The joists are placed broad-side upright to provide greater strength. In well-constructed homes, they are spaced 16 inches from center to center.

The bridging (FIG. 2-48) consists of small strips of 1×3-inch lumber, or a size near this, that are nailed diagonally between the floor joists along

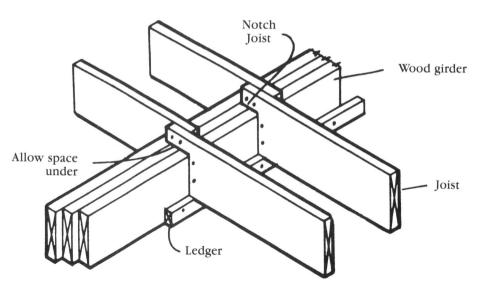

2-45 Typical girder and intertied joists.

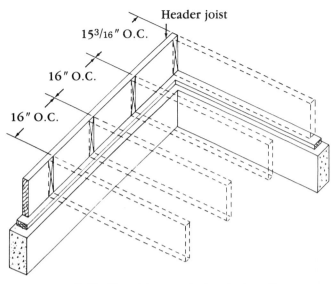

2-46   Floor joists attached to the box sill.

the center of the span. The purpose of the bridging is to keep the joists perpendicular so that they will provide the maximum amount of support and to distribute the weight on the floor between several joists rather than one or two. Bridging can also be made out of strips of metal.

The subfloor (FIG. 2-49) is the under flooring to which the finish floor is nailed. The subfloor is nailed directly to the floor joists and runs either

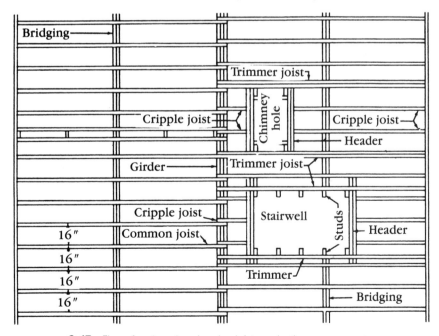

2-47   Floor framing plan showing joists and other components.

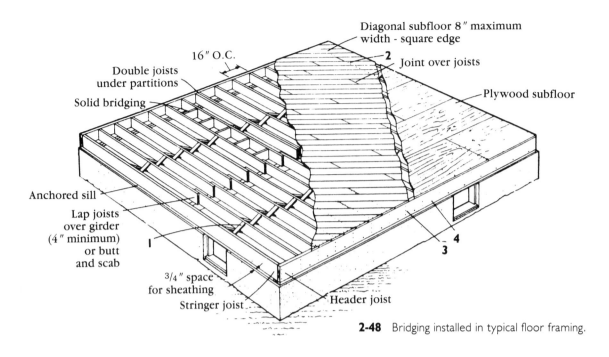

Diagonal subfloor 8″ maximum width - square edge

16″ O.C.

Double joists under partitions

Solid bridging

2

Joint over joists

Plywood subfloor

Anchored sill

Lap joists over girder (4″ minimum) or butt and scab

1

3

4

¾″ space for sheathing

Stringer joist

Header joist

2-48 Bridging installed in typical floor framing.

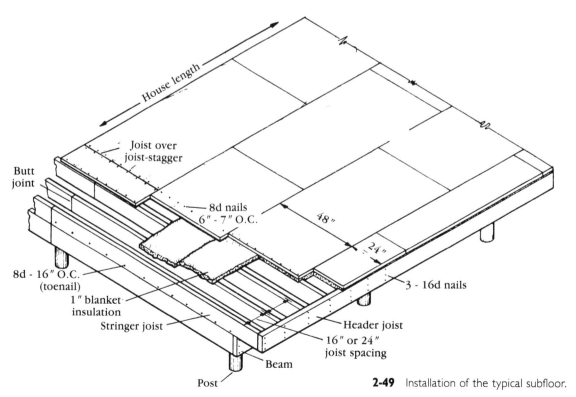

House length

Joist over joist-stagger

Butt joint

8d nails 6″ - 7″ O.C.

48″

24″

8d - 16″ O.C. (toenail)

1″ blanket insulation

Stringer joist

3 - 16d nails

Header joist

16″ or 24″ joist spacing

Post

Beam

2-49 Installation of the typical subfloor.

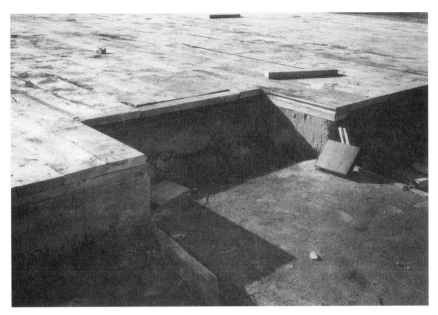

2-50 Actual installation of subfloor.

2-51 Close-up of subfloor installation.

at a 45- or 90-degree angle to the joists. The subfloor or rough floor not only furnishes a base for the finished floor, but also adds a degree of strength to the frame of the house. Figures 2-50 and 2-51 show an actual installation of subflooring.

An important part of energy efficiency in the home is the installation of adequate insulation under the flooring. Figure 2-52 illustrates how batt insulation is placed under flooring. This batt insulation is usually 48 inches long and 15 or 23 inches wide, depending on whether the joists are placed 16 or 24 inches apart.

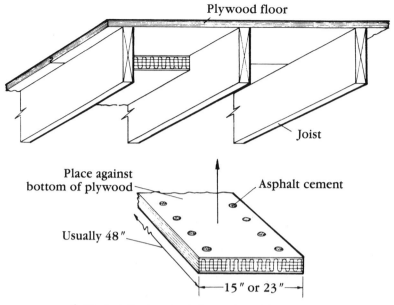

2-52 Installing batt insulation under subfloor.

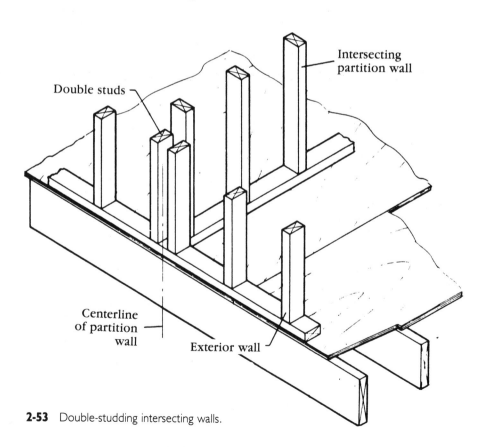

2-53 Double-studding intersecting walls.

Wall framing

Figure 2-53 illustrates how an intersecting partition wall is tied into the exterior wall with double studs. Notice that these plates and walls all rest on the subfloor which, in turn, rests on the joists.

Since you may be challenged by the installation of hardwood flooring on interior stairs, FIG. 2-54 shows a typical stairs installation and the many parts included. There are many different kinds of stairs, but all have two main parts in common: the treads people walk on and the stringers, which also are called strings, horses, and carriages, that support the treads. A very simple type of stairway, which consists only of stringers and treads, is shown in the left-hand view of the figure. Treads of the type shown here are called plank treads. This simple type of stairway is called a cleat stairway because of the cleats that are attached to the stringers to support the treads.

In a more finished type of stairway, the treads are mounted on two or more sawtooth-edged stringers and include risers, as shown in the right-

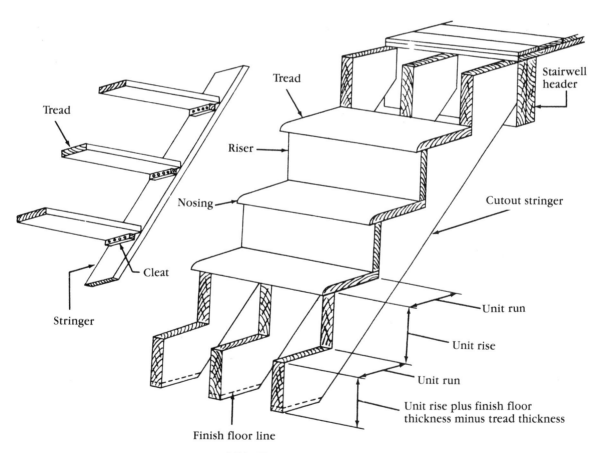

2-54 Typical stairs installation.

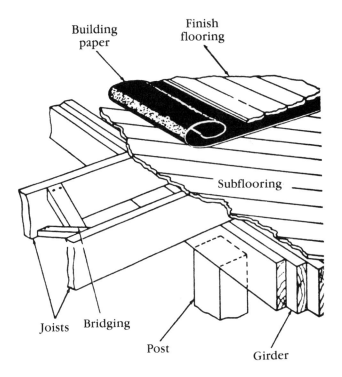

2-55   Installation of finish flooring over joists, bridging, girder, and subflooring.

hand view of FIG. 2-54. The stringers shown here are cut out of solid pieces of dimensional lumber (usually 2 × 12) and are therefore called cut-outs or stringers.

Flooring

Figure 2-55 illustrates how the finish flooring is added on to the house construction. Figure 2-56 illustrates how square-edged flooring is nailed and installed over the subflooring. First, building paper is installed on top of the subflooring. Then the finish flooring is installed at right angles and face-nailed.

Figure 2-57 shows the steps for installing tongue-and-groove finish flooring over a subfloor. Again, building paper can be applied between

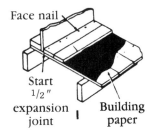

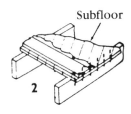

2-56   Nailing square-edged flooring.

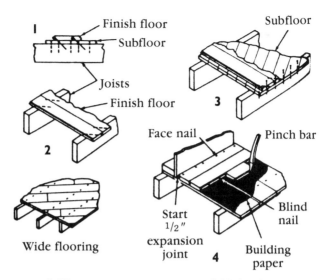

2-57 Nailing tongue-and-groove finish flooring.

the subfloor and the finish floor in order to keep moisture from coming up through the subfloor and damaging the flooring.

In the next chapter, you'll learn the specific steps for the installation of hardwood strip and block flooring in a variety of situations and using the numerous types of flooring that are available to the consumer. Remember that the planning of your hardwood flooring job is vital to its success.

Chapter **3**

Installing hardwood floors

The installation of hardwood flooring is, in one way, complex. The variety of types, sizes, and application systems is so great that there are literally dozens of ways to install hardwood floors. Once you've selected the type of flooring, however, the installation becomes much easier. Each type of hardwood flooring has its own installation methods, and manufacturers and dealers are usually helpful in offering instructions.

The purpose of this chapter is to guide you through the installation of strip, block, and other types of hardwood flooring so that you can do a professional-looking job—even if you've never worked with wood before. More than 50 illustrations will show you exactly how it's done.

HARDWOOD FLOOR INSTALLATION METHODS

Figure 3-1 illustrates the installation of wood-strip flooring over a concrete slab floor. Since this type of flooring tends to be cold and plain, many people install wood flooring over the top. First, be sure that there is sufficient insulation and a vapor barrier between the slab and the ground, or at least the slab and the flooring. The anchored sleepers in the illustration are intended to allow space between the cold slab and the hardwood floor. Insulation can be installed here, if you desire.

Another way of installing strip flooring over a concrete slab is offered in FIG. 3-2. This method allows a greater air pocket between the concrete and the flooring and includes a waterproof coating and vapor barrier to retard moisture, which is flooring's worst enemy. Nailing must be more precise in this method to make sure that the flooring is anchored to the sleepers. This can be accomplished by using a chalk line to mark the nailing line or by careful nailing.

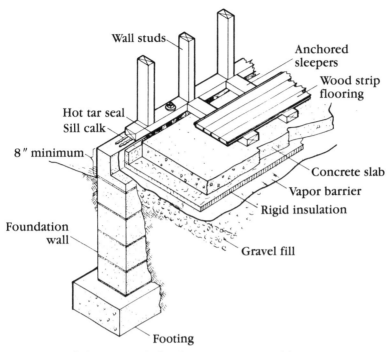

Wall studs

Anchored
sleepers

Wood strip
flooring

Hot tar seal
Sill calk

8″ minimum

Foundation
wall

Concrete slab

Vapor barrier

Rigid insulation

Gravel fill

Footing

3-1 Wood-strip flooring over a concrete slab.

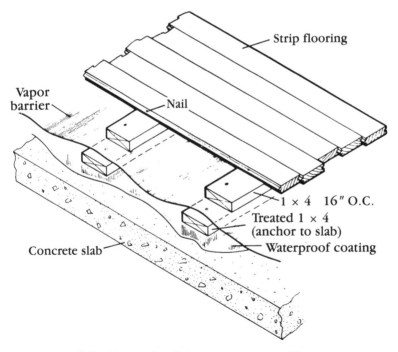

Strip flooring

Vapor
barrier

Nail

1 × 4 16″ O.C.

Treated 1 × 4
(anchor to slab)

Waterproof coating

Concrete slab

3-2 Alternate installation over a concrete slab.

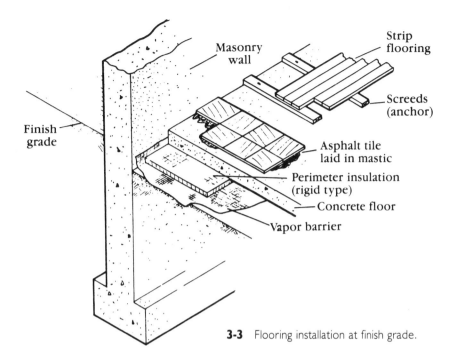

3-3 Flooring installation at finish grade.

In both cases, be sure that the treated lumber won't rot and eventually destroy your flooring from below.

Figures 3-3 and 3-4 illustrate the installation of flooring over a concrete slab at and below finish grade. I'll cover the installation of wood flooring over a concrete slab in more detail in the coming pages.

In most cases, strip and block flooring will be installed over a wood subflooring, as discussed in Chapter 2. The typical wood-subfloor system (FIG. 3-5) includes the floor joists, subfloor, building paper, and finish flooring. The flooring should be installed at a 45- or 90-degree angle (FIG. 3-6) to the subflooring.

MATERIALS HANDLING AND STORAGE

Most hardwood flooring has been kiln-dried to the proper moisture content. To maintain the moisture level, don't truck or unload it in rain, snow, or excessively humid conditions. Cover it with a tarpaulin or vinyl if the atmosphere is foggy or damp.

Check the job site before the flooring is delivered. Be sure that the flooring will not be exposed to high humidity or moisture. Surface drainage should direct water away from the building. Basements and crawl spaces must be dry and well-ventilated. In joist construction with no basement, provisions for outside cross ventilation must be provided through vents or other openings in the foundation walls. The total area of these openings should equal $1^1/2$ percent of the first floor area. A ground cover of 4- or 6-mil polyethylene film is essential as a moisture barrier.

3-4 Flooring installation below finish grade.

Flooring that is installed over a heating furnace or uninsulated ducts may develop cracks unless protection from the heat is provided. Use a double layer of 15-pound or a single layer of 30-pound asphalt felt or building paper, or 1/2-inch standard insulation board between the joists under the flooring in these areas. Insulation that is used over a heating plant should be nonflammable.

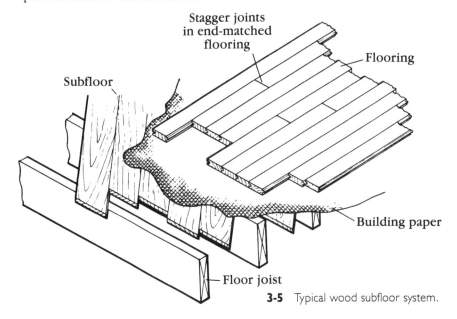

3-5 Typical wood subfloor system.

3-6 Installing strip flooring at a 90-degree angle to subflooring (courtesy Dixon Lumber Company, Inc.).

Before flooring is delivered, the building should be closed in, with outside windows and doors in place in a newly constructed home. All concrete, plaster, and other masonry should be thoroughly dry. In warm months, the building must be well-ventilated. In winter, a temperature of 65 to 70 degrees Fahrenheit (not higher) should be maintained at least 5 days before the flooring is delivered for best results.

When job conditions are satisfactory, have the flooring delivered, broken up into small lots, and stored in the rooms where it will be installed. Allow 2 to 3 days for the flooring to become acclimated to the job site. Such protection from heat, cold, and moisture extremes may seem wasteful, but they will pay off during the installation and maintenance of your hardwood floors.

INSTALLATION OVER CONCRETE SLABS

All types of hardwood flooring can be installed successfully over a concrete slab. The slab must be constructed properly, however, and the instructions below followed precisely.

Watch out for water. New concrete is heavy with moisture, an inherent enemy of wood. Proper on-grade slab construction requires a vapor barrier between the gravel fill and the slab. While this barrier prevents moisture from entering through the slab, it also retards the curing of the slab. So test for dryness, even if the slab has been in place over 2 years. To guide you, here are a few methods that professional builders and hardwood-flooring installers use to test concrete for moisture.

The rubber mat test   Lay a flat, noncorrugated rubber mat on the slab, place a weight on top to prevent moisture from escaping, and allow the mat to remain overnight. If there is trapped moisture in the concrete, the covered area will show water marks when the mat is removed. This test is only useful if the slab surface was originally light in color.

The polyethylene film test   Tape a 1-foot square of heavy, clear polyethylene film to the slab and seal all the edges with plastic packaging tape. If, after 24 hours, there is no clouding, or drops of moisture on the underside of the film, the slab can be considered dry enough to install wood floors.

The calcium chloride test   Place 1/4 teaspoon of dry (anhydrous) calcium chloride crystals, which are available at drug stores, inside a 3-inch diameter putty ring on the slab. Cover with a glass so that the crystals are totally sealed off from the air. If the crystals dissolve within 12 hours, the slab is too wet for a hardwood-flooring installation.

Keep in mind that the test should be made in several areas of each room and on both old and new slabs. The remedy for a moist slab is to wait until it dries naturally or to accelerate the drying process with heat and ventilation.

Vapor barrier

Start with a good vapor barrier. To be absolutely certain that moisture doesn't reach the finished floor, a vapor barrier must be used on top of the slab. Its placement depends on the type of nailing surface and/or the type of wood flooring to be used. Prepare the slab by sweeping it clean. The slab must be sound, level, and free from grease, oil stains, and dust. Level out any high spots and fill the low spots.

Plywood-on-slab method

This system uses 3/4-inch or thicker exterior plywood as the subfloor nailing base over the concrete slab. Refer to FIG. 3-7.

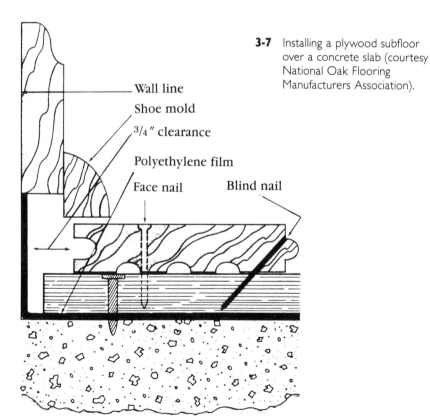

3-7 Installing a plywood subfloor over a concrete slab (courtesy National Oak Flooring Manufacturers Association).

Wall line

Shoe mold

3/4" clearance

Polyethylene film

Face nail   Blind nail

Roll out a 4-mil or heavier polyethylene film over the entire slab. Overlap the edges from 4 to 6 inches, allowing enough to extend under the baseboard on all sides. It does not need to be embedded in mastic.

Loosely lay plastic panels over the entire floor. Cut the first sheet of every run so that end joints will be staggered 4 feet. Level a 1/2-inch space at all the wall lines and allow about 1/8 to 1/4 space between panels.

Fasten the plywood with a powder-actuated concrete nailer or hammer-driven concrete nails. Use a minimum of nine nails per panel. Start at the center of the panel and work toward the edges, so that you are sure to flatten out the plywood and hold it securely.

An alternate method is to cut the plywood into 4×4-foot squares, score the back of each square, and then lay on mastic adhesive. This method, however, will require the use of either of two types of moisture barriers that can be laid in mastic. These will be described later in this chapter.

Screeds method

The screeds method of installing strip flooring on a concrete slab is shown in FIG. 3-8. This method uses flat, dry 2×4-inch screeds, or sleeps,

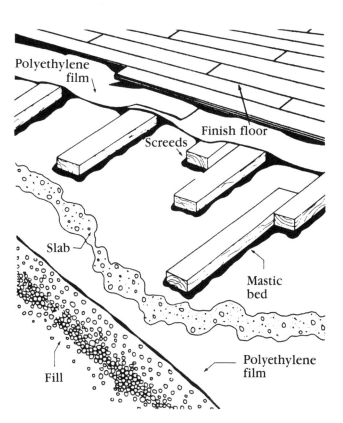

3-8 Screeds method of installing flooring over concrete slab (courtesy National Oak Flooring Manufacturers Association).

of random lengths from 18 to 48 inches. They must be preservative-treated with a product other than creosote, which might bleed through and stain the finished floor.

Sweep the floor clean, prime it with an asphalt primer, and allow it to dry. Apply hot, poured asphalt mastic and embed the screeds, 12 to 16 inches on center, at right angles to the direction of the finished floor. Stagger the joints and lap ends at least 4 inches. Leave a 3/4-inch space between the ends of the screeds and the walls.

Over the screeds spread a vapor barrier of 4- or 6-mil polyethylene film with the edges lapped 6 inches or more. It is not necessary to seal the edges or to affix the film with mastic, but avoid bunching or puncturing the plastic, especially between screeds. The finished flooring will be nailed through the film to the screeds.

Some installers prefer to use a two-membrane asphalt felt or building-paper vapor barrier, which will be explained later. The screeds are laid in rivers of mastic on the asphalt felt or building paper. In this system, the polyethylene film over the screeds is recommended because of the extra moisture protection that is provided at a nominal cost.

The screeds methods alone—that is, without a subfloor and spaced 12 inches on center—is satisfactory for all strip flooring and plank flooring to a 4-inch width. Plank flooring wider than 4 inches requires either the plywood-on-slab subfloor outlined earlier or screeds plus a wood subfloor in order to provide an adequate nailing surface. The subfloor may be 5/8-inch or thicker plywood or 3/4-inch boards.

Installing block flooring over a slab

Parquet, block, herringbone, and similar floors are normally laid in asphalt mastic and thus don't require a nailing surface on top of the slab. The need for a good moisture barrier is most important, however. The barrier can be achieved by either of the following methods.

Polyethylene method   Prime the slab with an asphalt primer and allow it to dry. Apply cold-type, cut-back asphalt mastic with a straightedged trowel to the entire slab surface. Allow it to dry for 30 minutes. Unroll a 4-mil polyethylene film over the slab, cover the entire area, and overlap the edges by 4 inches. Walk in the film by stepping on every square inch of the floor to ensure proper adhesion. Small bubbles are of no concern.

Two-membrane asphalt felt or building paper method   Prime and apply mastic with a notched trowel (FIG. 3-9) at the rate of 40 square feet per gallon. Let set 2 hours. Roll out 15-pound asphalt felt or building paper and overlap the edges by 4 inches. Butt the ends. Apply another coating of mastic with the notched trowel and roll out a second layer of asphalt felt for building paper. Lay both layers in the same direction, but stagger the overlaps to achieve a more even thickness.

The finish floor will be laid in mastic on the vapor barrier. This method applies only to tongue-and-groove parquet.

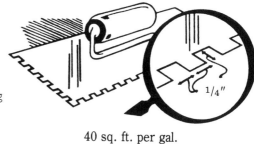

3-9 Notched trowel for applying mastic (courtesy Pennwood Products Co.).

40 sq. ft. per gal.

INSTALLATION OVER WOOD JOISTS

Let's move on to the installation of hardwood flooring over wood joists. Use exterior plywood or boards of No. 1 or No. 2 common pine or other softwood that is suitable for subfloors over wood joists. If you use plywood, it must be at least ¹/₂ inch thick. Lay the panels with the grain of the faces at right angles to the joists and nail every 6 inches along each joist. Use appropriate nails for the plywood thickness. Leave a ¹/₈-inch space between the panels.

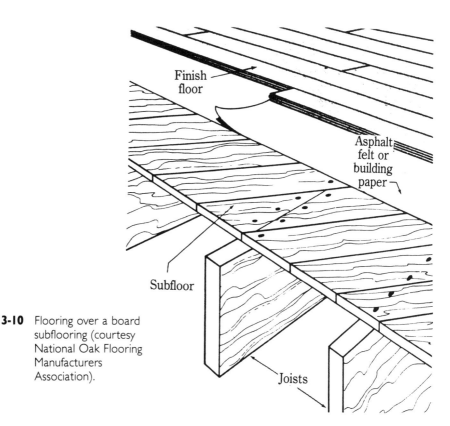

3-10 Flooring over a board subflooring (courtesy National Oak Flooring Manufacturers Association).

For a board subfloor (FIG. 3-10), use only flat, dry, 3/4-inch dressed square-edged boards that are no wider than 6 inches. Lay them diagonally across the joists with a 1/4-inch space between the boards to allow for expansion. Don't use tongue-and-groove boards. Nail to every bearing point with two 8d or 10d common nails. All butt joints must rest on bearings.

Mark the location of the joists so that the flooring can be nailed into them. If subfloor boards are used over sleepers or screeds, allow a 1/2-inch space between boards.

Good nailing is important. It keeps the board rigid and prevents creeping, which is sometimes caused by the shrinkage of the subfloor lumber. Without adequate subfloor nailing, it is impossible to obtain a solid, nonsqueaking floor.

FINISHED FLOOR LAYOUT

Here are instructions for applying strip flooring that is to be laid on plywood-on-slab, screeds, and plywood or board subfloors.

When a plywood or board subfloor is used, start by renailing any loose areas and sweeping the subfloor clean. Then cover it with a good grade of 15-pound asphalt felt or building paper and lap 4 inches at the seams to help keep out dust, retard moisture from below, and help prevent squeaks in dry seasons.

For the best appearance, lay the flooring in the direction of the longest dimension of the room or building—across or at right angles to the joists, as shown in FIG. 3-10. If a hallway parallels the long dimension of the room, begin the flooring by snapping a chalk line through the center of the hall and work from there into the room. Use a slip-tongue to reverse direction when you complete the hall later.

Location and straight alignment of the first course is important. Refer to FIGS. 3-11 through 3-13. Place a strip of flooring 3/4 inch from the starter

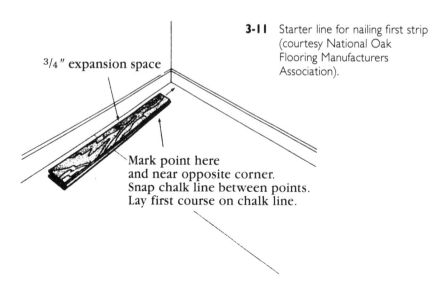

3/4" expansion space

3-11   Starter line for nailing first strip (courtesy National Oak Flooring Manufacturers Association).

Mark point here and near opposite corner. Snap chalk line between points. Lay first course on chalk line.

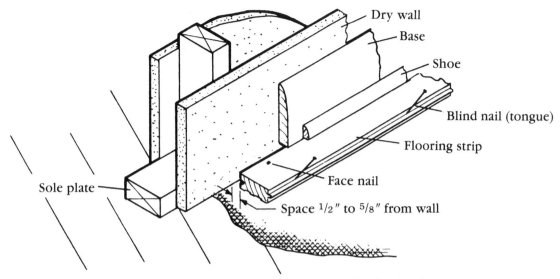

3-12 How initial strip will evenly fit into complete flooring picture.

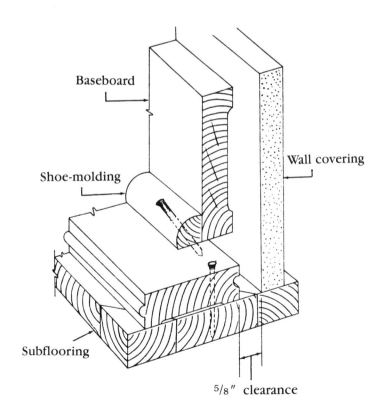

3-13 Cross section of initial strip and molding.

wall (or leave as much space as will be covered by the base and shoe molding) and put the groove side toward the hall. Mark a point on the subfloor at the edge of the flooring tongue. Do this near both corners of the room and then snap a chalk line between the two points. Nail the first strip with its tongue on this line. The gap between the strip and the wall is needed for expansion space and will be hidden by the shoe mold.

If you're working with screeds on the slab, you won't be able to snap a satisfactory chalk line on the loose polyethylene film that has been laid over the screeds. Make the same measurements and stretch a line between the nails at the wall edges. Remove the line after you get the starter board in place.

Lay the first strip along the starting chalk line, tongue out, and drive 8d finish nails at one end of the board near the grooved edge. Drive additional nails at each joist or screed and at the midpoints between joists. Keep the starter strip aligned with the chalk line. Predrilled nail holes will prevent splits. Nail additional boards in the same way to complete the first course.

Refer to FIGS. 3-14 through 3-18. They illustrate the steps for matching, inserting, nailing, and setting nails when you install tongue-and-groove hardwood flooring.

Next, lay out seven or eight loose rows of flooring end to end and in a staggered pattern with the end joints at least 6 inches apart. Find or cut pieces to fit within 1/2 inch of the end wall. Watch your pattern for an even distribution of long and short pieces and to avoid clusters of short boards.

Fit each board snug, groove to tongue, and blind-nail through the tongue according to TABLE 3-1. Refer to FIGS. 3-19 and 3-20 for methods of wedging floor boards tight.

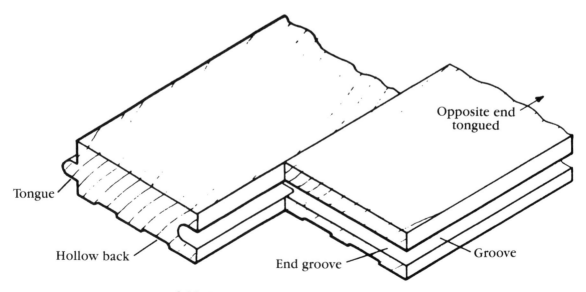

3-14 Matching tongue-and-groove flooring.

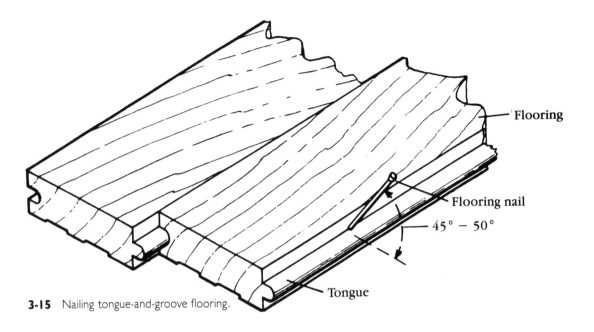

3-15 Nailing tongue-and-groove flooring.

Flooring

Flooring nail

45° − 50°

Tongue

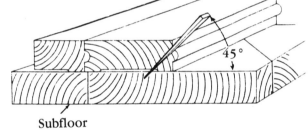

3-16 Toenailing a flooring nail into the subfloor.

45°

Subfloor

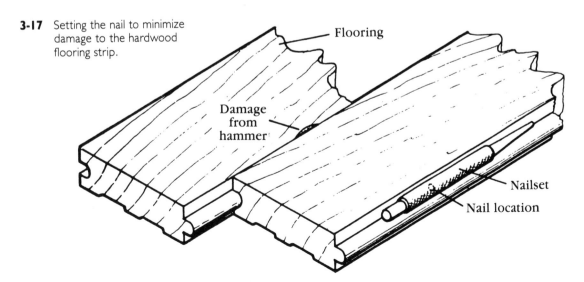

3-17 Setting the nail to minimize damage to the hardwood flooring strip.

Flooring

Damage from hammer

Nailset

Nail location

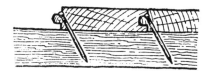

3-18   Cross section of toenailed flooring strips.

After the second or third course is in place, you can change from a hammer to a power nailer (FIG. 3-21), which is easier to use, does a much better job, and doesn't require countersinking. The power nailer drives a special barbed fastener, which is fed into the machine like a staple, through the tongue of the floor at the proper angle. Power nailers can be rented from many rental yards or hardwood-flooring suppliers.

When using the power nailer to fasten ³/₄-inch strip or plank flooring to plywood that has been laid on a slab, be sure to use a ³/₄-inch cleat. The usual 2-inch cleat may come out of the back of the plywood and prevent the nails from countersinking properly. In all other applications, the 2-inch cleat is preferred.

Table 3-1   Nailing Schedule

Flooring Nominal Size, Inches	*Size of Fasteners*	*Spacing of Fasteners*
³/₄ × 1¹/₂	2″ machine driven fasteners; 7d or 8d screw or cut nail.	10″–12″* apart
³/₄ × 2¹/₄	,,	
³/₄ × 3¹/₄	,,	
³/₄ × 3″ to 8″ plank	,,	8″ apart into and between joists.
Following flooring must be laid on a subfloor.		
¹/₂ × 1¹/₂ ¹/₂ × 2	1¹/₂″ machine driven fastener; 5d screw, cut steel or wire casing nail.	10″ apart
³/₈ × 1¹/₂ ³/₈ × 2	1¹/₄″ machine driven fastener, or 4d bright wire casing nail.	8″ apart
Square-edge flooring as follows, face-nailed—through top face.		
⁵/₁₆ × 1¹/₂ ⁵/₁₆ × 2	1″, 15-gauge fully barbed flooring brad. 2 nails every 7 inches.	
⁵/₁₆ × 1¹/₃	1″, 15-gauge fully barbed flooring brad. 1 nail every 5 inches on alternate sides of strip.	

*If subfloor is ¹/₂ inch plywood, fasten into each joist, with additional fastening between joists.

National Oak Flooring Manufacturers Association

3-19 Using a pinch bar and wedge to tighten flooring strips.

Continue across the room. End up on the far wall with the same ³/₄-inch space allowed on the beginning wall. It may be necessary to rip a strip to fit.

Avoid nailing into a subfloor joint. If the subfloor joint is at right angles to the finish floor, don't let the ends of the finish floor meet over it.

When nailing directly to screeds (no solid subfloor), nail at all screed intersections and to both screeds where a strip passes over a lapped screed joint. Since flooring ends are tongue-and-groove, all end joints do not need to meet over the screeds. End joints of adjacent strips, however, should not break over the same void between screeds.

Some long boards may have horizontal bends or sweeps that have resulted from a change in moisture content. A simple lever device (FIG.

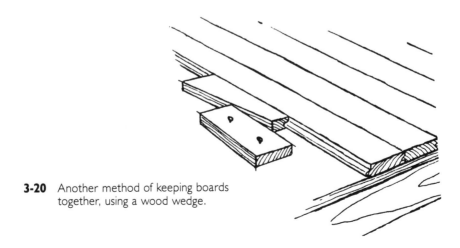

3-20 Another method of keeping boards together, using a wood wedge.

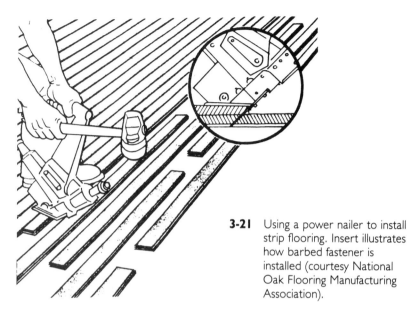

3-21 Using a power nailer to install strip flooring. Insert illustrates how barbed fastener is installed (courtesy National Oak Flooring Manufacturing Association).

3-22) can be made on the job to force such boards into position, as well as to pull up several courses. Once the entire floor is in place, nail the shoe molding to the baseboard, not to the flooring.

PLANK FLOORING

Plank flooring is normally made in 3- to 8-inch widths and may have countersunk holes for securing the planks with wood screws (FIG. 3-23). These holes are then filled with wood plugs, which are supplied with the flooring in many cases. Plank flooring is installed in the same manner as strip flooring by alternating courses by widths. Start with the narrowest boards, then the next width, etc., and repeat the pattern.

Manufacturers' instructions for fastening the flooring vary and should be followed. The general practice is to blind-nail through the tongue as with conventional strip flooring and then to countersink one or more (depending on the width of the plank) No. 9 or No. 12 screws at each end of each plank and at intervals along the plank to hold it securely. Cover

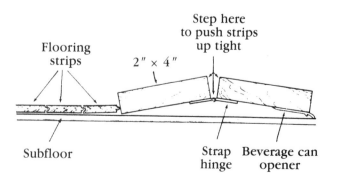

3-22 Job-made lever used for forcing strip flooring boards up tight (courtesy National Oak Flooring Manufacturers Association).

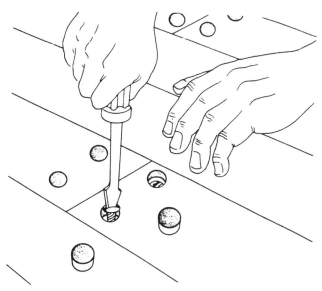

3-23 Installing wood screws in countersunk holes drilled in plank flooring. Plugs are then inserted over screws (courtesy National Oak Flooring Manufacturers Association).

the screws with wood plugs that have been glued into the holes. Take care not to use too many screws because with the plugs in place, they tend to give the flooring a polka-dot appearance.

Be sure the screws are the right length. Use 1 inch if the floor is laid over ³/4-inch plywood on a slab. Use 1 to 1¹/4 inches in wood-joist construction or over screeds.

Some manufacturers recommend face nailing in addition to other fasteners. Another practice sometimes recommended is to leave a slight crack, about the thickness of a putty knife, between planks.

LAYING OVER OLD FLOORING

You can also lay a new strip floor over an old floor. In this case, the existing wood floor can serve as a subfloor.

Drive down any raised nails, renail loose boards, and replace any warped boards that can't be leveled. Sweep and clean the floor well, but don't use water. Remove the thresholds to allow the new flooring to run flush through the doorways. Remove doors and baseboards.

Lay asphalt felt or building paper over the old floor, as discussed earlier. Always install the new floor at right angles to the old floor boards. This method is sometimes the best way to handle an old hardwood floor that is in need of extensive repair.

PARQUET AND BLOCK FLOORING

The styles and types of block and parquet flooring, as well as the recommended procedures for their application, vary somewhat among the different manufacturers. Detailed installation instructions are usually provided with the flooring or are available from the manufacturer or distributor.

I'll cover the installation of parquet, block, herringbone, and similar flooring here and later in this chapter. This section applies only to tongue-and-groove parquet flooring where tongues and grooves are engaged. It doesn't apply to slat-type or finger-block parquet.

First, lay both blocks and the individual pieces of parquet flooring in mastic to a wood subfloor or over a moisture barrier, as described earlier. Use a cold, cutback asphalt mastic and spread it over the entire area that is to be floored at the rate of 1 gallon per 40 square feet. Use the notched edge of the trowel. Allow the mastic to harden a minimum of 2 hours or up to 48 hours, as directed by the manufacturer. The surface will be solid enough after 12 hours to allow you to snap working lines on it. Use blocks of the flooring as stepping stones to snap lines and begin the installation.

There are two ways to lay out parquet flooring. The most common is with edges of parquet units, and thus the lines they form, square with the walls of the room. The other way is a diagonal pattern , with the lines at a 45-degree angle to the walls.

Let's learn about the installation of the square pattern first. Never use the walls as a starting line because walls are almost never truly straight. Instead, use a chalk line to snap a starting line about 3 feet or so from the handiest entry door to the room and roughly parallel to the nearest wall. Place this line exactly equal to four or five of the parquet units from the center of the entry doorway.

Next find the center point of this baseline and snap another line at an exact 90-degree angle to it from wall to wall. This will become your test line to help keep your pattern straight as the installation proceeds. A quick test for squareness is to measure 4 feet along one line from where they

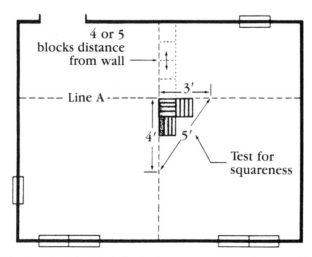

3-24 Working lines for laying block flooring in a square pattern (courtesy National Oak Flooring Manufacturers Association).

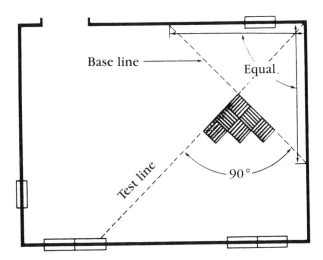

3-25 Working lines for laying block flooring in a diagonal pattern (courtesy National Oak Flooring Manufacturers Association).

intersect, and 3 feet along the other. The distance between these two points will be 5 feet if the lines are true (FIG. 3-24).

Here's how to lay a diagonal pattern of parquet block flooring. Measure equal distances from one corner of a room along both walls and snap a chalk line between these two points to form the baseline. This pattern need not be at a precise 45-degree angle to the walls in order to appear perfect. A test line should again intersect the center of the baseline at an exact 90-degree angle (FIG. 3-25).

Special patterns also are easy to install. Most exciting parquet patterns can be laid out with the two working lines that were just covered. Herringbone will require two test lines, however. One will be the 90-degree line already discussed. The other line should cross the same intersection of lines, but at a 45-degree angle to both (FIG. 3-25).

If such elaborate, preliminary layout preparation seems a bit overdone, keep in mind that it is wood you are installing. Each piece must be carefully aligned with all its neighbors. Small variations in size, which are natural to wood, must be accommodated during installation to keep the overall pattern squared up. You cannot correct a creeping pattern after it develops. A carefully laid-out floor causes less problems.

Wood parquet must always be installed in a pyramid or stairstep sequence, rather than in rows. This method again prevents the small inaccuracies of size that are found in all wood flooring from magnifying or creeping until the pieces look misaligned.

Place the first parquet unit carefully at the intersection of the base and test lines. Lay the next units along the line ahead and to the right of the first ones. Then continue the stair-step sequence. Carefully watch the corner alignment of new units with those that are already in place. Install in a quadrant of the room. Leave the trimming at the walls until later. Then return to the base and test lines and lay another quadrant, again

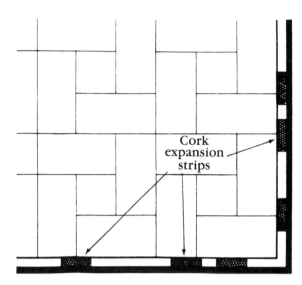

Cork
expansion
strips

3-26   Use of cork blocking around edges of a block floor (courtesy National Oak Flooring Manufacturers Association).

repeating the stair-step sequence. Cover this area with the flooring. Work
backwards from the baseline toward the door. A reducer strip may be
needed at the doorway.

Most wood floor mastics, regardless of the type or open time, will
allow the tiles to slip or skid when sidewise pressure is applied for some
time after the open time has elapsed. By working from knee boards or
plywood panels that have been laid on top of the installed area of flooring, you can avoid this sidewise pressure. For the same reason, no heavy
furniture or activity should be allowed on the finished parquet floor for
about 24 hours. Some mastics must also be rolled.

Cut blocks or parquet flooring pieces to fit at the walls. Allow a
³/₄-inch expansion space on all sides. Use cork blocking in 3-inch lengths
between the flooring edge and the wall to permit the flooring to expand
and contract (FIG. 3-26).

With blocks, a diagonal pattern is recommended in corridors and in
rooms where the length is more than 1¹/₂ times the width. This diagonal
placement minimizes expansion under high humidity conditions.

SOLVING SPECIAL INSTALLATION PROBLEMS

Every do-it-yourselfer would prefer typical installations that have no
handicaps or problems to overcome. Life doesn't seem to work this way,
however. Here are instructions on how to solve special hardwood-
flooring installation problems.

Oak flooring over a radiant-heated concrete slab   Flooring will not impair the efficiency of the heating system, but slightly higher water temperatures may be required. An outside thermostat is therefore recommended to anticipate rapid temperature changes. Boiler water temperature must be controlled to keep it to a maximum of 125 degrees Fahrenheit and so limit the temperature of the slab surface to about 85 degrees, which is an acceptable level for most mastics.

The flooring should be installed as in any other slab project, except do not fasten the plywood to the concrete with either nails or powder-actuated fasteners. Turn on the heating system at least 48 hours before the flooring is delivered to the job because the heat will drive remaining moisture out of the slab. Allow the flooring to become acclimated to the environment for 2 or 3 days, then install by the recommended slab methods covered earlier in this chapter. Remember to check the flooring and mastic manufacturers' specifications for suitability for use over radiant heat.

Strip flooring in a wood plenum system   This method of housing construction utilizes a crawl space that is completely sealed to the outside as a plenum to which air from the heating/cooling system is supplied. The air then enters each room through floor ducts.

A ground cover of polyethylene film is essential. The heating system must also operate for at least 48 hours prior to the delivery of the flooring in order to stabilize the moisture condition. No other special considerations are necessary to install the flooring.

FITNESS ROOM FLOORING

Hardwood flooring is popular for school gyms, athletic clubs, and home fitness rooms. Gymnasium floor products are often made of 3/4-inch pecan or maple. Beach and oak are also suitable. It is most important to have some resiliency built into these floors. In most respects, however, installation closely follows the screeds-in-mastic method, which is recommended for conventional use, with a plywood or board subfloor installed over the screeds.

Make sure the slab is dry and level with a good float finish. Maximum surface variation is 1/4 inch in 10 feet. Grind down high areas and fill low areas with a concrete leveling compound.

Sweep the slab clean and prime with asphalt primer. Let dry thoroughly and coat with asphalt mastic. Use a notched trowel and apply at a rate of 40 square feet per gallon. Embed a layer of 15-pound asphalt felt or building paper. Start at a wall with a half sheet and lap seams. Cover this with another layer of mastic and embed a second layer of asphalt felt or building paper. Start at the same wall with a full sheet to cover the seams of the first layer.

Either hot or cold mastic is satisfactory. If the cold type is used, be sure to allow 2 hours for the solvents to evaporate before you apply the building paper.

An alternate method of surface damp-proofing is to embed a 6-mil polyethylene film in a cold mastic, as described earlier. Lap the film edges 6 inches.

A suspended concrete slab needs no surface damp-proofing. Cross-ventilation below the slab is essential, however, and if the slab is suspended over exposed earth, a ground covering of 6-mil polyethylene should be provided.

Screeds used in this application are identical to that previously described, with the following exceptions. Place them on 12-inch centers. If a subfloor is used, 16-inch centers are allowed. Leave a 2-inch space between the ends of the screeds and the base plate on all walls to allow for expansion.

The finish flooring may be nailed directly to the screeds. A much more sound and satisfactory floor can be achieved, however, by installing a subfloor of a minimum of 3/4-inch plywood or 3/4-inch dressed square-edged boards that are no wider than 6 inches.

Follow the arrangement and nailing schedules described previously. If boards are used, leave a 1/2-inch space between them.

Start laying the finish flooring in the middle of the room and work toward the walls. Place the first two courses groove to groove and use a slip-tongue joint to join the strips. Face-nail as well as blind-nail both courses. Proceed with succeeding courses in the conventional manner. Use either 7d or 8d cut steel nails, screw-type nails, or 2-inch barbed fasteners.

After an area 3 to 4 feet wide has been laid across the room, leave a 1/16-inch expansion space between the last course laid and the next course (FIG. 3-27). Repeat the expansion space evenly at 3- to 4-foot intervals across the room.

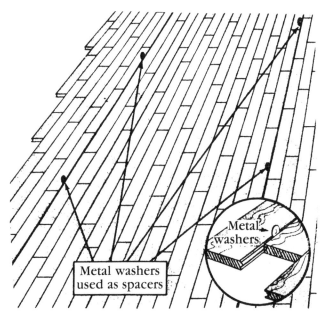

3-27 Use of metal washers to provide expansion space on a gym floor (courtesy National Oak Flooring Manufacturers Association).

Nailing is most important. Nail to all screeds and to both screeds when a strip passes over a lapped screed joint. All end joints do not need to meet over screeds, but adjoining strips should not break over the same screed space. If a subfloor is used, nails must be no more than 10 to 12 inches apart.

Allow a 2-inch expansion space along all walls and doorways. The space can be covered at the walls with an angle iron that is bolted to the wall or a special wood molding, and at the doorways by a metal plate that is designed for such use.

MAKING INSTALLATION EASIER

The National Oak Flooring Manufacturers' Association recommends a number of tips for easier and better flooring installation.

Work from left to right. When laying strip flooring, you'll find it easier to work from your left to your right. Left is determined when your back is to the wall where the starting course is laid. When it is necessary to cut a strip to fit to the right wall, use a strip that is long enough for the cutoff piece to be 8 inches or longer. Start the next course on the left wall with this piece.

Save short pieces for closets. For best appearances, always use long flooring strips at entrances and doorways. Save some of the short pieces for closet areas and scatter the rest evenly in the general floor area.

Put a frame around an obstruction. You can give a much more professional and finished look to a strip-flooring installation if you frame hearths and other obstructions and use mitered joints at the corners.

Reverse the direction of strip flooring. Sometimes it's necessary to reverse the direction of the flooring to extend it into a closet or hallway. To do this, join the groove edges, with a slip-tongue that is available from flooring distributors. Nail it in place in the conventional manner.

Use only sound, straight boards for subfloors. The quality of the subflooring will affect the finish flooring. Use only square-edged, 3/4-inch dressed boards that are no wider than 6 inches. Boards that have been used for concrete form work are often warped and damp and should not be used.

Don't pour concrete after the flooring is installed. Concrete basement floors are sometimes poured after hardwood flooring has been installed. Many gallons of water from the drying concrete evaporate into the atmosphere of the house, however, where it may be absorbed by hardwood flooring and other wood components. This is not a recommended building practice since excessive moisture will cause problems with wood floors and other woodwork. Wood flooring should not be installed until after all concrete and plaster work is completed and dry.

Put the voids between the screeds to good use. Masonry insulation fill, which is normally used in hollow concrete blocks, can be poured between the screeds or sleepers of a slab installation to give additional moisture protection and deaden the drumming sound that sometimes occurs from foot traffic.

Deaden sound in a multi-story building. Noise transmission from an upper to a lower floor can be reduced in the following way. Nail the subfloor to the joists in the normal manner and then cover it with 1/2-inch or thicker cork or insulation board that has been laid in mastic. Cover this with another 3/4-inch plywood subfloor, which has been laid in mastic. Nail the finish strip or plank floor to the plywood, or lay block or parquet flooring in mastic on the plywood. In the case of parquet, the second subfloor plywood can be a 1/2-inch tongue-and-groove type. Note that specifications for some high-rise apartment buildings call for other types of sound-deadening construction and should be followed.

INSTALLING A FLOATING FLOOR

As mentioned earlier, there are numerous ways of installing a satisfactory hardwood floor. Thus far, in this chapter you've learned about the more traditional methods. Another method is called the floating floor system.

Figures 3-28 and 3-29 illustrate a cross section of a floating floor installation over wood for suspended concrete with a foam underlayment. If the subfloor is in contact with the ground, or on-grade, a 6-mil polyethylene film barrier should first be installed over the subfloor and the foam applied over the poly film (FIGS. 3-30 and 3-31). Lap the seams of poly film at least 8 inches. If the subfloor is not level to within 3/16 inch in a 10-foot radius, it should be leveled with a latex leveling compound.

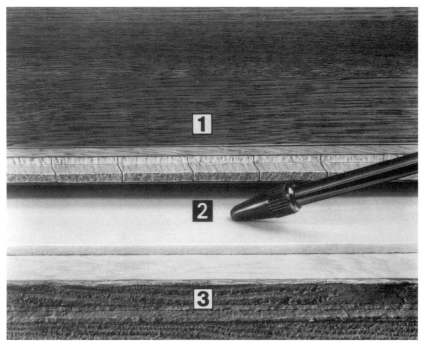

3-28   Cross section of floating floor installation of (1) plank, (2) foam underlayment, and (3) plywood subfloor (courtesy Harris-Tarkett, Inc.).

Total finished floor
elevation = $5/8''$

Foam   Longstrip

3-29 Floating floor over wood or
suspended concrete (courtesy
Harris-Tarkett, Inc.).

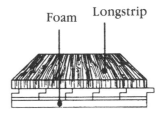

Wood or suspended concrete

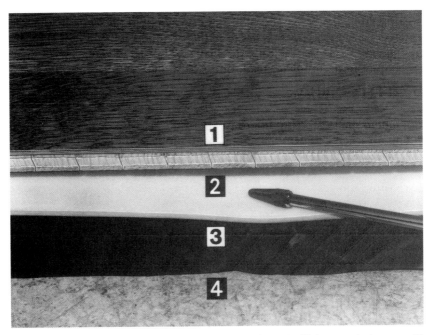

3-30 Cross section of floating floor installation of (1) plank, (2) foam underlayment, (3)
poly film, and (4) concrete (courtesy Harris-Tarkett, Inc.).

Poly seams lapped 8″
Foam underlay seams butted

6 mil poly film   Longstrip
Foam

3-31 Floating floor over concrete
on-grade (courtesy Harris-
Tarkett, Inc.).

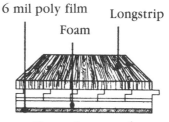

Concrete on-grade

Tools required for a floating installation include a hammer, a hand or power saw, wood, plastic, or equivalent 1/2-inch spaced wedges, crowbar, a chalk line, installation adhesive, a tapping cover block, and any special tools.

Once the poly film, if it is used, and the foam underlayment have been installed over the subfloor, the job site is ready for the boards. Don't open the bundles until you are ready to begin the installation process. Decide which direction the boards will run. Start at one sidewall (FIG. 3-32) with the first row of boards. Allow a 1/2-inch expansion along the side and end walls (FIG. 3-33) by using wood or plastic wedges, or equivalent spacers. If the starting wall is out of square, it is recommended that the first row of boards be scribed to allow for 1/2 inch of expansion (FIG. 3-34) and a straight working line.

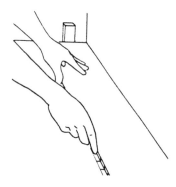

3-32 Start at one sidewall (courtesy Harris-Tarkett, Inc.).

3-33 Use wedges to establish expansion space (courtesy Harris-Tarkett, Inc.).

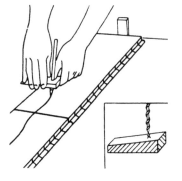

3-34 Scribing boards (courtesy Harris-Tarkett, Inc.).

3-35 Applying adhesive along side grooves (courtesy Harris-Tarkett, Inc.).

3-36 Applying adhesive on end joint (courtesy Harris-Tarkett, Inc.).

The specially milled boards must be edge- and end-glued using the manufacturers' adhesive or equivalent. Apply the adhesive in 8-inch long beads with a 12-inch space between the beads (FIG. 3-35) along the side grooves. Fully glue every end joint (FIG. 3-36). If any excess glue squeezes out on to the finish surface, wipe it off with a moist cloth.

Install the first row. Use the appropriate expansion space and face the grooved side toward the wall (FIG. 3-37). The subsequent rows are installed and edge- and end-glued with a hammer (FIG. 3-38) and tapping block to prevent damage to the protruding tongue. Check for a tight fit on the sides and ends. Stagger 2 feet between the end joints of adjacent board rows (FIG. 3-39).

Most often, the last row does not fit in width. When this occurs, follow this simple procedure. Lay a row of unglued boards with the tongued side toward the wall, directly on top of the last installed row (FIG. 3-40). Take a short piece of the strip and put it against the wall. Moving down the wall, draw a line with a pencil along the row. The resulting line gives the proper width for the last row which, when cut (FIG. 3-41), can then be wedged into place using a crowbar and cover board (FIG. 3-42) to prevent damage to finished walls or molding.

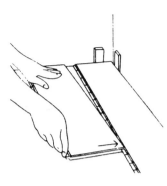

3-37   Installing first row (courtesy Harris-Tarkett, Inc.).

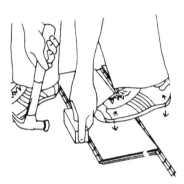

3-38   Installing subsequent rows with a hammer and block (courtesy Harris-Tarkett, Inc.).

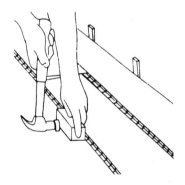

3-39   Check for tight fit of boards (courtesy Harris-Tarkett, Inc.).

3-40   To make last row fit in width: first, lay a row of boards on top of the last installed row (courtesy Harris-Tarkett, Inc.).

Parquet hardwood flooring offers a rich look to a modern home.

Plank hardwood flooring offers a traditional look.

Darker hardwood stains accent darker wood furniture.

Hardwood flooring stands up to heavy foot traffic in a mall.

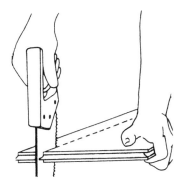

3-41 Cut the board on the scribed line (courtesy Harris-Tarkett, Inc.).

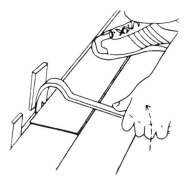

3-42 Wedge the last row of boards into place with a crow bar (courtesy Harris-Tarkett, Inc.).

3-43 Attach trim to the wall, not the flooring (courtesy Harris-Tarkett, Inc.).

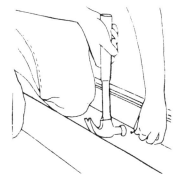

Make sure that wedges or spacers are removed when the installation is complete. The expansion space should also be covered with an appropriate molding that allows the boards to move freely underneath. Always attach the trim to the wall or vertical object (FIG. 3-43) and never to the floor boards.

One more note: in large areas that measure more than 24 linear feet, use a ¼-inch expansion for each 12-linear feet of width and length.

INSTALLING PLANK FLOORING

General installation methods for plank flooring were covered earlier in this chapter. Here's another popular method.

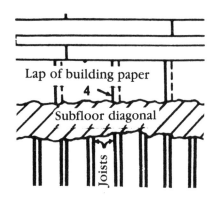

3-44   Cross section of a plank-flooring installation (courtesy Harris-Tarkett, Inc.).

   Suggested tools and accessories include a conventional claw hammer or power nailer and a No. 80, 3/4-inch counterbore, unless the flooring is prebored at the factory. You'll also need 7d, 2¹/₄-inch, screw-type flooring nails for conventional nailing or power cleats for power nailing. No. 9, 1¹/₄-inch flathead wood screws and walnut or oak wooden plugs, ⁵/₁₆ × ³/₄-inch in diameter, should be used.

   Figure 3-44 illustrates a cross section of a plank flooring installation, including the subflooring. Kiln-dried coniferous lumber, which is 1×4 inch to 6 inch wide, square-edged, and laid diagonally over 16-inch, on-center wood joists, is recommended. A minimum of ¹/₂-inch thick exterior plywood can also be used with the long edges at right angles to the joists and staggered so that the end joists on adjacent panels break over different joists. Particleboard is not considered a suitable subfloor.

   Refer to FIGS. 3-45 through 3-48. Lay out your plank flooring only after the plasterboard and tile work have thoroughly dried and all but the final woodwork and trim have been completed. The building interior should have been dried and seasoned and a comfortable working temperature should exist during the plank flooring installation.

   Make sure that the subfloor is adequate and properly nailed before you start. Clean the subfloor surface and cover it with 15- or 30-pound asphalt saturated felt. Lap the edges at least 4 inches. Double the felt around the heat ducts in the floor.

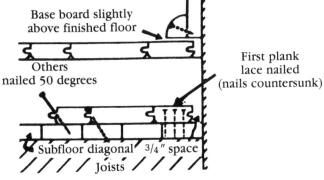

3-45   Installation details of plank flooring (courtesy Harris-Tarkett, Inc.).

3-46 Using spacers between plank flooring (courtesy Harris-Tarkett, Inc.).

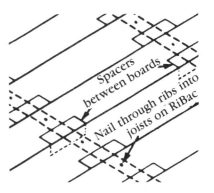

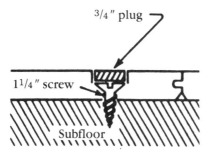

3/4″ plug

1¹/4″ screw

Subfloor

3-47 Installation of wood screw and plug to fasten plank flooring (courtesy Harris-Tarkett, Inc.).

3-48 The finished plank floor (courtesy Harris-Tarkett, Inc.).

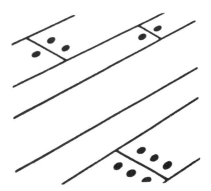

Plank flooring should be laid at right angles to the floor joists and, if possible, in the direction of the longest dimension of the room. Begin laying tongue-and-groove plank flooring in a room corner with the edge groove of the planks facing the wall. Provide no less than a 3/4-inch expansion space or what will be covered by the baseboard and trim.

The first run of planks should be face-nailed and then countersunk. All other runs should be nailed at a 50-degree angle on 8-inch centers at the tongue. Make sure the end joints are staggered. Remember to leave a small space between the planks—usually 1/32 inch is sufficient.

After nail installation of the plank flooring, bore appropriate holes for screws and plugs. Use the counterbore if the flooring was not factory bored or if additional boring is desired. Screws and plugs should be located at the ends of each plank and at intervals frequent enough to hold the planks securely. Whenever possible, screws should extend into the joists. The number of screws to use is a function of job conditions and the desired surface appearance. Use one screw and plug on each end of 3- and 4-inch wide planks, two widthwise on the ends of 5- and 6-inch planks, and three widthwise on the ends of 7- and 8-inch planks.

How to sand and finish your plank flooring will be covered in Chapter 4.

INSTALLING PARQUET FLOORING

One of the easiest and most beautiful floorings you can install is the parquet floor (FIG. 3-49). Early parquet floors were extremely difficult to install correctly, but modern manufacturing and application techniques have placed parquet flooring within the can-do realm of the typical do-it-yourselfer.

Prepare the subfloor in the same manner as for other types of hardwood flooring. Make sure it's nailed down tightly, clean, and level. If you've stripped an old floor off to install your parquet flooring, make sure the mastic and fasteners are gone. Old wood should be thoroughly sanded with $3^1/2$ open-grit paper to remove oils, waxes, paints, varnishes, glues, and other foreign matter. After sanding, the subfloor should be swept and vacuumed to remove dirt. An alternate method is to cover the existing subfloor with $1/2$-inch plywood.

Figure 3-50 shows the proper method of laying out the room prior to installation and the proper procedure to follow in laying the parquet blocks. Locate the center of the room by measuring from the center of the length and width sides with a chalk line. Then determine the number of parquet blocks that are to be laid from the center of the room to the wall.

3-49 Parquet block flooring is actually easy to install (courtesy Pennwood Products Co.).

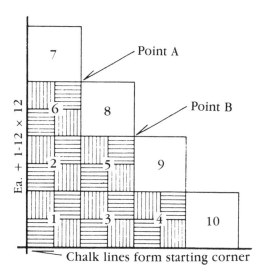

3-50 Laying out a parquet block floor (courtesy Pennwood Products Co.).

If the measurement leaves less than 2 inches of block next to the wall, shift the centerline to allow 2 inches or more of the block to be laid at the wall. Be sure that at least $1/2$ inch or more of an expansion space is left around all four walls. Following installation, cover the expansion area with baseboard or quarter round.

Spread the adhesive according to label instructions. Lay a quadrant of the room from the centerline with blocks. Complete that area before you start the next. The accurate alignment of the first 10 blocks is most important. Never force or tap blocks with a hammer. Just tightly lift the tongue-and-groove blocks by hand. Be sure that corners form right angles and match properly at points A and B in FIG. 3-50.

ADHESIVE

Adhesive is used in most hardwood flooring installations. It is especially critical to the installation of parquet block flooring.

Spread the adhesive on the floor with a notched trowel. Apply the adhesive to an area of such size that the flooring can be laid into the adhesive within approximately 2 hours. Don't let a heavy skin form on the adhesive. A light skin on the adhesive can easily be broken by pressing the flooring into it.

After you spread the adhesive, wait to install the flooring. Leave the adhesive open to help ensure a faster bond. Allow 20 to 30 minutes for bonding, depending on the temperature and humidity.

To install the flooring, press down firmly. Lift a piece of flooring occasionally to be sure the adhesive is transferring to the block. Remember that wood flooring needs expansion space around its edges.

Once the installation is complete, some adhesives require that you roll the floor carefully in both directions with a 75 to 100 pound roller.

To remove adhesive from tools, use naptha, white gas, or mineral spirits. Be careful not to get it on a surface it could damage.

Your hardwood floor is installed! Congratulations. The work isn't over, though. The final step in the installation of a hardwood floor is the finishing, which includes sanding and then applying one of the many types of finishes that will bring out the wood's natural beauty while protecting it from wear.

Chapter **4**

Finishing hardwood floors

While some hardwood flooring can be pur-
chased prefinished, the majority must be sanded and finished once the
installation is completed. In this chapter, you will learn how to finish your
new hardwood floor using the methods and equipment that have been
developed by professional hardwood flooring installers over many years.
Chapter 6 will cover the refinishing of old hardwood floors.

Finishing your hardwood floor requires some specialized and expensive power tools. All can be rented, however, at most rental yards or from
the larger flooring-supply houses. Some suppliers will even loan you
equipment free or at a nominal charge if you purchase the majority of
your flooring needs through them. These primary pieces of floor-finish-
ing equipment include the drum sander, the power edger, and the floor
polisher. Finish application equipment can also be rented or borrowed.

PREPARING THE FLOOR

Apply the finish to the hardwood floor should be one of the last jobs of
any construction project. In this way, other work and the traffic from
workers won't mar the finish. Wall coverings should be in place and the
paint completed except for a final cost on the base molding. Sweep the
floor clean immediately before sanding (FIG. 4-1).

SANDING

A drum-type floor sander (FIG. 4-2) is used for heavy sanding operations. A
floor polisher with a sanding or screen disc or steel wool is used for special situations and to give the floor an extremely fine finish. Professional
floor finishers will also use a spinner-type edger in areas where the drum
sander can't reach. Since the spinner-type edger is difficult to use, how-

4-1   This 50-year-old hardwood floor can be refinished to look like new.

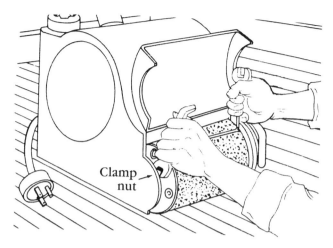

4-2 Drum-type floor sander (courtesy National Oak Flooring Manufacturers Association).

ever, you may elect to hand-scrape and hand-sand around the edges of the room.

Load the drum sander with a medium abrasive (FIG. 4-3). Place the machine along the right-hand wall with about ⅔ of the length of the floor in front of you. Start the motor and ease the drum to the floor. Walk slowly forward, letting the machine pull you along at an even pace. As you near the wall, gradually raise the drum off the floor by lifting up on the control handle.

Start pulling the machine backward and ease the drum to the floor. Cover the same path as you made on the forward cut. When you reach your starting point, ease the drum from the floor and move the machine to one side by approximately 4 inches. Then repeat the forward and back-

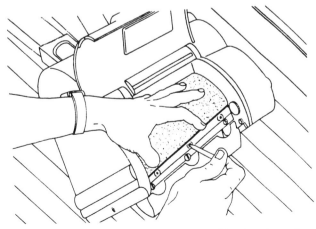

4-3 Another type of drum sander has the paper held in place by a clamp (courtesy National Oak Flooring Manufacturers Association).

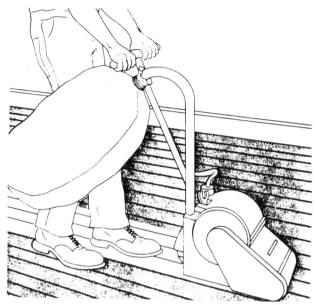

4-4 Always keep the drum sander moving when it is operating (courtesy National Oak Flooring Manufacturers Association).

ward passes. When ²/₃ of the room has been sanded, turn the machine in the opposite direction and sand the remaining third in the same manner. Be sure the cuts made on the last ¹/₃ of the room overlap the first cuts by 2 to 3 feet. This method blends the two areas together.

It is very important that you never let the sanding drum touch the floor unless you are moving the machine forward or backward (FIG. 4-4). If you do, it will cut a hollow in the floor that cannot be removed.

After you finish the first cut with the drum sander (FIG. 4-5), use the power edger (FIG. 4-6) or a hand-scraper and sand up to the baseboard, in

4-5 Finishing the first cut with the drum sander (courtesy National Oak Flooring Manufacturers Association).

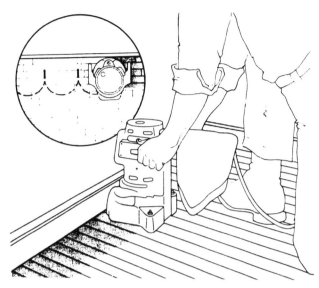

4-6 Use the power edger to sand up to the baseboard and in other areas the drum sander won't reach (courtesy National Oak Flooring Manufacturers Association).

corners, closets, and other areas where the drum sander won't reach. Use the same grit as used on the drum sander. Move the edger in a brisk, left-to-right semicircular pattern.

When using a hand scraper (FIG. 4-7) instead of the power edger, apply even pressure and scrape in the direction of the grain or from the

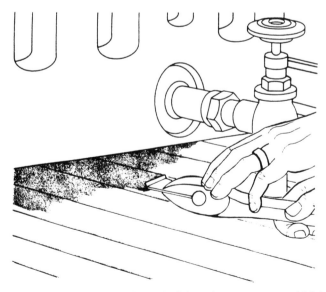

4-7 A hand scraper may be used instead of the edger to remove old finish in tight spots. The hand scrapper is recommended for the inexperienced (courtesy National Oak Flooring Manufacturers Association).

wall into the sanded area if the floor is parquet. Avoid gouging the wood with the scraper. After scraping, use a sanding block and paper with the same grit as that used on the drum sander to smooth the flooring. A brick with a piece of old blanket glued around it makes a good sanding block.

After you have sanded the entire floor with a medium abrasive, repeat the entire procedure with a fine abrasive (TABLE 4-1). Sand the body of the floor first and then the edges. Pay particular attention to blending the edges with the main floor area.

Often only two sanding cuts are used, but for a smoother, finer finish, switch from the drum sander to a floor polisher that has been fitted with fine paper or a screen disc and sand the entire floor. If the floor is to be stained, however, use a slightly heavier grit on this cut to leave a tooth on the wood surface, which will enable the stain to penetrate more readily.

Sanding strip and plank flooring

If the floor is flat and level, make all the cuts parallel to the direction of the strips (FIG. 4-8). If the floor is uneven, however, make the first cut at a 45-degree angle to the direction of the strips. This will remove any peaks or valleys that are caused by minute variations in the thickness of the strips or in the subfloor. Make succeeding cuts parallel to the direction of the strips. Always use at least two cuts. A third and even a fourth cut with the floor polishing machine and sanding disc is recommended because of the fine finish it imparts to the floor.

Table 4-1   Sanding New Floors

Floor	Operation		Grade of Sandpaper	
Hardwood		Uneven Floor	Medium-Coarse	2 (36)
Oak, Maple,	First Cut	Ordinary Floor	Fine	1 (50)
Beech, Birch	Final Sanding		Extra Fine	2/0 (100)
Softwood Pine	First Cut	Uneven Floor	Medium-Fine	1 1/2 (40)
		Ordinary Floor	Fine	1 (50)
Fir	Final Sanding		Extra Fine	2/0 (100)

4-8 Sand in the longest direction of the room (courtesy National Oak Flooring Manufacturers Association).

Sanding parquet, block, and similar flooring

Use the drum sander for the first two cuts. Make the first cut with a medium-grit paper on a diagonal of the room and the second cut with a fine-grit paper on the opposite diagonal. Then switch to the floor polishing machine and, using a screen disc, make a third lengthwise cut following the longest room dimension (FIG. 4-9).

PREPARING FOR THE FINISH

When the sanding is completed, sweep or vacuum the floor clean (FIG. 4-10). Wipe up all the dust on windows, sills, doors, door frames, base-

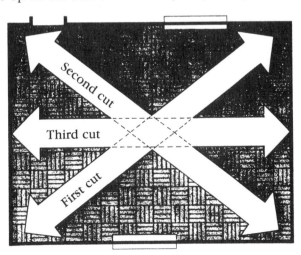

4-9 Parquet, block, herringbone, and similar flooring should have three sanding cuts, as shown (courtesy National Oak Flooring Manufacturers Association).

4-10   The sanded hardwood floor.

boards, and the floor using a painter's tack rag.

Inspect the floor carefully and hand-sand to remove any scratches or swirl marks. Fill the cracks, spaces, and nail holes with a commercial putty of a matching color, which is compatible with stain and/or finish, or make your own putty with fine dust from the final sanding and a floor sealer. Mix them to form a thick paste. Apply the putty with a putty knife and scrape off the excess. When it has dried, hand-sand it with a fine-grit paper. When these operations are finished, both old and new floors become essentially new wood surfaces and should be treated as new floors.

Apply the first coat of stain or other finish the same day that the sanding is completed to help keep the wood grain from rising and creating a rough surface. With stain and all finishing materials, be sure to read and follow the label instructions. Avoid vigorous shaking or stirring, which may cause air bubbles to be trapped in the material and affect the quality of the finish. Allow adequate ventilation and avoid breathing of the fumes for prolonged periods.

Finishing your hardwood floor involves numerous products, such as a sealer, stains, shellac, and varnish (FIG. 4-11) and application equipment (FIGS. 4-12 through 4-15). To guide you, TABLE 4-2 suggests numerous types of finishes for the specific types of wood that are used in home construction.

TYPES OF FINISHES

Let's consider the properties and the methods of application for the more popular finishes that are used on hardwood floors: penetrating sealers, surface finishes, polyurethane, varnish, shellac, lacquer, and bleaching.

4-11 Popular hardwood flooring finishes include varnish, sealer, and shellac (courtesy National Oak Flooring Manufacturers Association).

4-12 Hardwood-flooring finishes are often applied using a paint brush (courtesy National Oak Flooring Manufacturers Association).

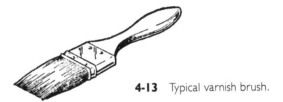

4-13 Typical varnish brush.

4-14 Two large flat brushes.

Penetrating sealers

This is the finish that is recommended for most residential floors. The sealer soaks into the wood pores and hardens to seal the floor against dirt and most stains. It wears only as the wood wears and will not chip or

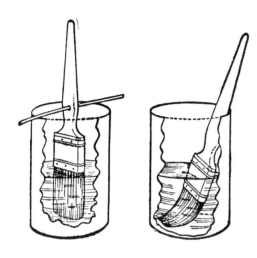

4-15 Right and wrong way to clean brushes in solvent.

Table 4-2   Home Interior Woods

Type	*Characteristics*
American Woods	
Hard:	
Birch	Takes all kinds of stains well; no filler required; used for veneers, to imitate other woods, for trim, interior finishings, furniture; good for blonde finishes
Cherry	Takes light reddish stain; generally no filler needed unless lacquered; will not bleach; used for furniture, to imitate mahogany
Hickory	Takes water stain best; takes fine polish after sanding; generally a heavy filler needed; used for furniture, to imitate mahogany and walnut
Maple	Stains well in lighter colors; takes fine finish; no filler needed; widely used for furniture, veneers, floors
Oak	Takes many kinds of stains; heavy filler generally needed, but not for many period effects; widely used, especially for floors and heavier pieces of furniture
Walnut	Takes many stains well; bleaches well; heavy filler generally needed, except for old English effect; used for furniture, veneers
Soft:	
Cypress	Takes oil and water stains well; close, even grain; used for interior and exterior trim, garden furniture
Fir	Takes oil stains well; has very strong figure; used for interior trim; doors, flooring, also shelves, cupboards

Table 4-2 Continued.

Type	Characteristics
Pine, yellow	Takes oil stain well; prominent grain; resinous; widely used for bedroom and kitchen furniture, closets, bookcases, floors
Pine, white	Takes oil and water stains well; no filler needed generally; straight grained; when finished is generally used to imitate maple
Redwood	Takes red stain well; does not readily bleach or take lighter stains; filler sometimes used; used for veneers
Imported Woods	
Avodire	Natural finish preferable; takes light stain; requires filler; bleaches well
Circassian Walnut	Stains well; bleaches well; requires filler; has very fancy figure; used mostly for veneers
Mahogany, Philippine	Takes all stains well; needs heavy filler; has very beautiful grain; used for furniture, trim, floors, to imitate walnut and true mahogany; may fade if not properly finished
Mahogany, true	Takes all stains well; generally needs filler; excellent color and figure; used for furniture; often called the finest of all woods
Rosewood	Requires "washing" before finishing; needs filler; will not bleach; used mostly for veneers
Satinwood	Takes light stains; often finished naturally; needs filler; will not bleach; used mostly for veneer and inlay

scratch. After years of wear, the floor can usually be refinished without sanding by cleaning it and applying another coat of sealer or a special reconditioning product. Limited areas of wear can be refinished without showing lap marks where new finish is applied over the old.

Penetrating sealers can also be used as an undercoat for varnish or shellac when a high gloss is desired. They are available in natural, a number of wood tones, and other colors.

There are two basic types of sealers, which are distinguished by their drying time requirements. Normal or slow-drying sealers can be used safely by anyone. Fast-drying sealers should be used only by a professional who is accustomed to handling and applying them and can complete the job within the allotted drying time to avoid lap marks or a splotchy appearance. Some sealers produce satisfactory results with one coat, but most manufacturers recommend two coats or one plus a special top dressing.

Make sure that the space between boards has been filled before applying a sealer (FIG. 4-16). Then brush sealer around door edges (FIG. 4-17). Sealers are mopped on the general floor area. Use a clean string mop or long-handled applicator with a lamb's wool pad (FIG. 4-18). Secondary

4-16   Older hardward floors may require wood filler to fill between boards that have dried with age.

choices for applicators are a wide brush or squeegee. Generously apply the sealer and wipe up the excess with clean cloths or a squeegee.

Note   Some manufacturers call for the sealer to be rubbed into the wood with steel wool pads on an electric buffer while the sealer is still wet. Figures 4-19 through 4-22 illustrate the application of penetrating sealer.

4-17   Use a small brush to apply finish around the edges of the floor.

4-18 A long-handled lamb's wool applicator is excellent for applying penetrating sealer finishes.

4-19 Apply the finish in the same direction as the wood's grain.

4-20 Work your way through the room and into the next room, making sure you have an exit when you are done.

Allow the sealer to dry and then buff the floor with No. 2 steel wool by hand or use an electric polisher that is equipped with a steel wool pad (FIG. 4-23). For even greater gloss and maximum protection from wear, use a single application of a penetrating sealer and then buff with steel wool. Follow this with two coats of polyurethane, the application of which will be covered later.

4-21 The finished floor will have a bright sheen as it shows off the wood's natural grain and character.

Stains

If other than a natural finish is desired, coloring must be the first step in the finishing process. Let's first look at coloring sealers and stains.

Colored penetrating sealers These are generally available in a number of wood tones and several other colors. They are an easy finishing method since the floor is colored as part of the operation. Their application will be covered later.

Pigmented wiping stains These are penetrating, oil-resin products to which pigments have been added. The pigments are not in solution, but in suspension, so the material must be stirred regularly during use to maintain a uniform color. The pigment collects in the open pores of the wood and accentuates the grain pattern.

Varnish stains These are a formulation of varnish with an oil-soluble dye that provides the color. As with penetrating sealers, you color the wood

4-22   The finished hardwood floor.

at the same time you apply the finish coat. The main disadvantage to varnish stains is that they tend to obscure the wood grain to a great extent and thus do not provide nearly as beautiful a finish as one of the other methods.

Stains are applied by brushing on a generous coat. The pigment is allowed to penetrate the wood and then it is wiped off with clean rags. Wipe vigorously to remove as much stain as possible from the surface.

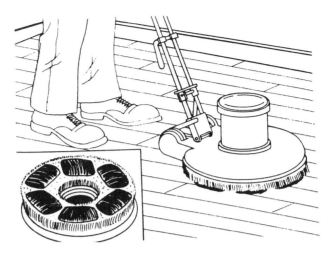

4-23 Ordinary steel wool pads may be used with the floor polisher when buffing a sealer finish (courtesy National Oak Flooring Manufacturers Association).

The pigment that remains in the pores and grain lines will provide the color. Always test the stain on an extra board because the length of time the stain is allowed to remain on the floor will determine the degree of color tone provided.

Surface finishes

All of these finishes are applied with a high-quality brush or lamb's wool applicator that is designated for use with the particular type of finish selected.

On strip and plank flooring, work in the direction of the boards in a path narrow enough to keep a wet working edge as you move across the room (FIG. 4-24). With block and other patterned floors, apply the finish in strips the width of one block, or about 12 inches for other types. Work from wall to wall across the shorter room dimensions.

Lap strokes by working from the wettest area back into the area that was just covered. Follow the manufacturer's directions for drying times between coats.

Polyurethanes

Finish manufacturers have done much to improve polyurethanes in the past few years. These blends of synthetic resins, plasticizers, and other film-forming agents produce an extremely durable surface that is moisture-resistant. They are the best choice for kitchens, commercial applications, and other areas that are exposed to stains, spills, and high traffic.

There are two types: oil-modified and moisture-cured. The first type of polyurethanes are air-drying materials—the solvents pass off in vapors.

4-24   Work in the direction of the boards (courtesy National Oak Flooring Manufacturers Association).

Because of the toxic fumes, oil-modified polyurethanes should not be applied by the nonprofessional.

Moisture-cured polyurethanes absorb small quantities of moisture from the air, which causes them to cure and harden. While slightly more wear-resistant than the air-drying type, the moisture-cured polyurethanes are extremely difficult to apply properly. Both the relative humidity or amount of moisture in the air, and drying time are crucial. Moisture-cured polyurethanes are thinner than oil-modified ones so that they typically take more coats to complete. On the plus side, they usually don't discolor with age so they can be touched up without concern for matching tones.

Somewhat similar to polyurethanes are the epoxy and ureaformaldehyde finishes, which are also synthetic products with high durability. The type of undercoat, working time, number of coats, and other factors are all critical and make application difficult. They should only be applied by the highly skilled.

Caution   The adhesion of polyurethanes, epoxies, and ureaformaldehyde finishes is affected by wax and grease, as well as some types of stains, bleaches, or sealers. Further more, one type of polyurethane may not be compatible with another type. Always make a test in a closet or some other inconspicuous place to be sure that the finish will adhere and dry properly. This is particularly important when you are refinishing an old floor, since some of the old finish may have penetrated the wood fibers below the level to which it has been sanded.

Use a brush or lamb's wool applicator to apply the polyurethanes and work along the grain to get a smooth, even coat. Allow the drying time as specified by the manufacturer, then buff with steel wool. Dust thoroughly and apply a second coat along the grain of wood. The final coat does not require buffing.

For a particularly good finish that combines the best qualities of a sealer and a surface coating, use a penetrating sealer followed by one or more coats of polyurethane. Check the product labels to be sure that the sealer and top coat are compatible.

Varnishes

Use only varnish manufactured for flooring applications, since it is more durable than ordinary varnish. Varnish gives a glossy finish that has good durability and is resistant to stains and spots. It shows scratches, however, and worn spots are difficult to patch without showing lines between the old and new finishes.

Varnish drying time is 8 hours or more. Use three coats over bare wood, or two coats over a wood filler or a shellac or sealer undercoat.

Thin a clear varnish with one part thinner to eight parts of varnish for the first coat over bare wood. Use full-strength varnish over penetrating a stain, shellac, or colored varnish. Apply all succeeding coats full strength and observe the drying times recommended by the manufacturer. Remember that varnish takes much longer to dry than shellac or lacquer.

Flow varnish on as smoothly as possible and brush out each stroke. Sand with a fine grit sandpaper between coats. Dust well and wipe with a turpentine-dampened rag between coats. Use two or three coats.

Shellacs

Shellac is easy to apply and dries fast. Two or three coats can be applied in one day. It is moderately resistant to water and other staining agents, but will spot if liquids are not wiped up promptly. It gives a high gloss and will not darken with age as quickly as varnish. Use at least two coats and preferably three. Shellac may be used as a sealer under a varnish top coat.

The main disadvantage to shellac is that it chips easily. Repairs can be made by sanding the chipped or worn area and then touching it up with new shellac. Be careful to feather the new finish into the surrounding area.

Use a 3-pound cut—3 pounds of shellac dissolved in 1 gallon of denatured alcohol. Apply with a wide brush to get a full, smooth flow and avoid puddling. Allow it to dry 2 hours and then sand with fine paper. Dust and recoat. This time, allow 3 hours drying time before you apply a third coat. Sand with a fine grit paper, then dust before the third coat. Use extra-fine paper to smooth the final coat. Allow the shellac to harden overnight before you use the floor.

Lacquer

Lacquer provides a glossy finish with about the same durability as varnish. An advantage to lacquer is that worn spots may be retouched, since the new lacquer dissolves the old and blends with it rather than forming an additional layer. Lacquer is difficult to apply because it dries so quickly, which causes lap marks. The use of lacquer by individuals other than skilled applicators is not recommended. Two coats are required.

Apply lacquer with a wide brush or mohair roller. Work fast to prevent lap marks. Allow the first coat to dry 1 hour, then hand-sand with a fine grit paper and dust. Apply the second coat and allow it to dry overnight before you use the floor. Do not sand the final coat. For a three-coat job, allow an hour for the second coat to dry, then sand with a fine grit paper, dust, and apply a third coat that has been reduced with one part thinner to four parts lacquer. Be sure to use a thinner that is recommended by the lacquer manufacturer to avoid drying problems.

Bleaching

An interesting effect can be obtained on both new and old floors by bleaching the wood to remove the color without obscuring the grain pattern. The general tone of the wood is retained, but the color intensity is reduced.

Wood bleaches are available at most paint stores. The best are usually two-solution products. The first solution is brushed on and changes the pigments chemically, while the second solution removes the pigments. Before using a bleach, be sure that the flooring is clean and free from oils, grease, old finish, and any dirt that might repel the bleach, prevent it from working, or give it an uneven effect.

Since bleaches have a water base, their use will cause the grain of the flooring to rise. Sand the floor with a fine grit paper after using the bleach to restore the floor to the required smooth surface.

Follow manufacturer's instructions for application times on the bleach you use. It's a good idea to make a test on a piece of scrap flooring because the length of time the bleach remains on the floor will affect the amount of color that is removed.

A polyurethane finish may be applied to the bleached floor. Some finishers prefer, however, to follow the bleaching operation with a white stain or a wood filler that has been mixed with wood pigment, which gives the whitest floor of all. Then they use polyurethane for the final finish. Check the product label to be sure it is compatible with the stain or pigmented filler you want to use and that the polyurethane will not turn an amber color over the whitened floor.

Fillers

Most hardwoods have minute crevices that are exposed by sawing and sanding. Before the development of modern finishing products, it was common to use a wood filler to fill these crevices or pores. Penetrating sealers and polyurethane finishes do not require their use, and fillers are quite often omitted with other finishing materials as well.

There are, however, two situations when the use of a filler is helpful.

One is when a mirror-smooth finish is desired, usually with varnish as the top coat. By filling the pores of the wood, you get an absolutely smooth surface that, with a gloss varnish, has an extremely high light reflectance.

The other use for fillers is when a floor is colored. In this case, a pigmented filler of the desired color tone is applied. This process is identical to bleaching, except that the filler is tinted instead of white.

Apply the filler with a 4-inch, short-bristled, flat brush. Cover a small area at a time. Brush with the grain first, then across the grain. When it dulls over, but before it hardens, wipe it vigorously with burlap or other coarse rags. Wipe across the grain first, then with the grain. Move on to another area until all the flooring is filled. Let the filler dry for 24 hours and then disc sand with a fine grit paper before you apply other finish materials.

STENCILING

Decorative boarders or overall floor patterns can be applied to hardwood floors with the use of stencils and masking tape, which are available at many paint and decorating stores. The stencils or tape must be securely adhered to the finished floor surface to prevent the final color from bleeding beyond the area that is to be treated. Refer to FIG. 4-25.

One method uses a clear or light-colored penetrating sealer over the entire floor area and a darker color sealer to create the pattern. For a more vivid color and pattern, use a good-quality floor enamel. Always make a

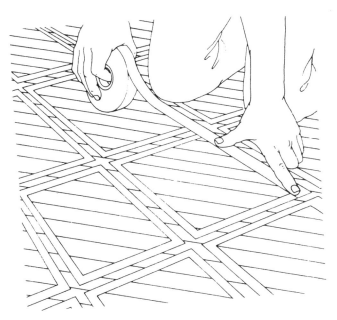

4-25 Many interesting effects may be obtained by stenciling (courtesy National Oak Flooring Manufacturers Association).

test panel first to be sure you get the desired effect. If you use paint, it will adhere more satisfactorily to the base finish.

Use paint sparingly and in other than high-traffic areas because it will chip and scratch easily. It should always be protected with one of the surface finishes described earlier to help prevent such damage.

GYM AND ROLLER-RINK FLOORS

The sanding and other initial preparations for applying a finish to gym floors are identical to those for floors in other installations. Special finishing products are made to provide the gloss and slip- and abrasion-resistant characteristics that are required, as well as to permit the painting of the game lines. Use only products that are manufactured for such applications and follow the specific finishing schedule that is provided since this varies from one product to another. Deviations from the recommended procedure may effect the quality and performance of the finish.

For those who are installing gymnasium or game floors of hardwood, Appendix B gives the specific markings and measurements for most popular gym games.

PROTECTING THE FLOOR'S FINISH

For the final touch of beauty and to protect the finish, apply one or more coats of good wax. Use either a liquid buffing wax/cleaner or paste wax. Use only brands that are designated for hardwood floors and, if a liquid, be sure it has a solvent, not a water base.

Apply the wax after the finish coat is thoroughly dry and polish it with a machine buffer. The wax will give a lustrous sheen to the floor and form a protective film that will prevent dirt from penetrating the finish.

Caution   Some manufacturers of urethane finishes do not recommend waxing, especially for commercial jobs, because the wax may make the floor slippery. Gymnasium and roller-rink floors should never be waxed. They require special maintenance products and procedures that are available from several manufacturers that also produce the finishing materials for such installations.

PROTECTING BASE MOLDINGS

Base moldings serve as a finish between the finished floor and wall (FIGS. 4-26 and 4-27). They are available in several widths and forms. The simplest is made by butt-jointing two pieces of wood at an inside corner (FIG. 4-28).

A two-piece base consists of a baseboard that is topped with a small base cap (FIG. 4-29A). When the plaster is not straight and true or a hardwood floor needs an expansion gap, the small base molding will more closely conform to the variations than will the wider base alone. A common size for this type of baseboard is $5/8 \times 3^{1}/4$ inches or wider.

A one-piece base varies in size from $7/16 \times 2^{1}/4$ inches to $1/2 \times 3^{1}/4$ inches and wider (FIGS. 4-29B and 4-29C). Although a wood member is

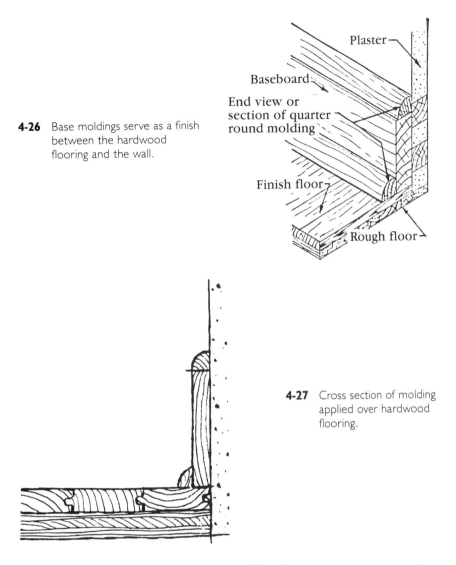

4-26 Base moldings serve as a finish between the hardwood flooring and the wall.

Plaster

Baseboard

End view or section of quarter round molding

Finish floor

Rough floor

4-27 Cross section of molding applied over hardwood flooring.

desirable at the junction of the wall and flooring to serve as a protective bumper, wood trim is sometimes eliminated entirely.

Most baseboards are finished with a base shoe that is $1/2 \times 3/4$ inches in size. A single-base molding without the shoe is sometimes placed at the wall-floor junction.

Square-edged baseboard should be installed with a butt joint at the inside corners and a mitered joint at the outside corners (FIG. 4-30). It should be nailed to each stud with two 8d finishing nails. Molded single-piece bases, base moldings, and base shoes should have a coped joint at the inside corners and a mitered joint at the outside corners.

A coped joint is one in which the first piece is square-cut against the plaster or base and the second piece is coped. It is accomplished by saw-

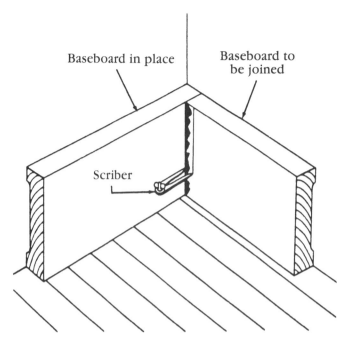

4-28 Butt-joint baseboard.

ing a 45-degree miter cut and then trimming the molding along the inner line of the miter with the coping saw (FIG. 4-31).

The base shoe should be nailed into the subfloor with long, slender nails and not into the baseboard itself or the hardwood flooring. Thus, if there is a small amount of movement in the hardwood flooring, the molding will allow this movement while covering it.

To guide you in the selection and installation of common molding types, let's look at some of the more popular designs (FIG. 4-32).

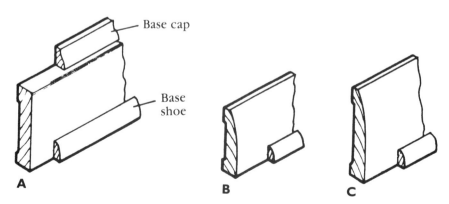

4-29 Baseboard topped with a small base cap (A), a narrow ranch base (B), and wide ranch base (C).

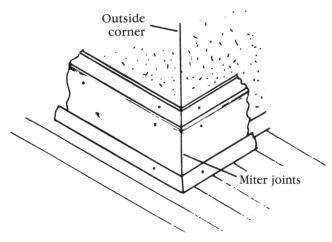

4-30 Milter joints at outside corner molding.

The term contour cutting refers to the cutting or ornamental face curves on stock that is to be used for molding or other trim. Most contour cutting is done on the shaper, which is equipped with a cutter or blades, or with a combination of cutters and/or blades that are arranged to produce the desired contour.

The simple molding shapes, which are shown in FIG. 4-32 are the quarter round, the half round, the scotia or cove, the cyma recta, and the cyma reversa. The quarter and half round form convex curves. The cove

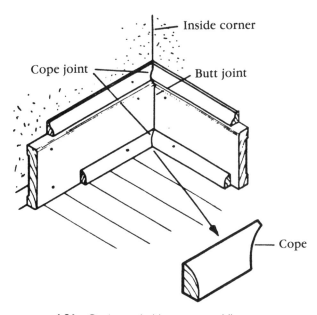

4-31 Coping an inside corner molding.

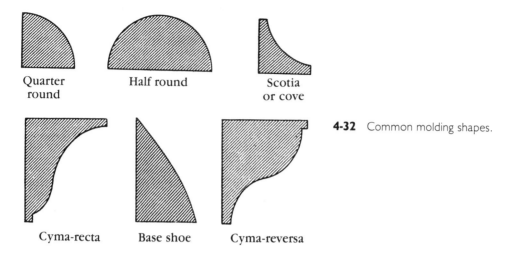

4-32 Common molding shapes.

molding forms a concave curve. And the cyma moldings are combinations of convex and concave curves.

Inside corner joints between the molding trim members are usually made by butting the end of one member to fit the face of the other member.

First, saw the end of the abutting member square, as you would for any ordinary butt joint between ordinary flat-faced members. Then miter the end to 45 degrees, as shown in the first and second views of FIG. 4-33. Set the coping saw at the top of the line of the miter cut, hold the saw at 90 degrees to the lengthwise axis of the piece, and saw off the segment shown in the third view. Closely follow the face line left by the 45-degree

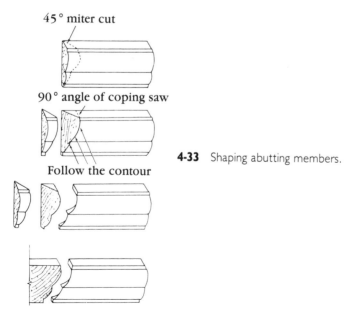

4-33 Shaping abutting members.

miter cut. The end of the abutting member will then match the face of the member, as shown in the third view. This is called a coping joint.

You've now completed the finishing of your hardwood floor. The floor has been prepared, sanded, and a finish installed. The molding around the edge has been cut and installed. All that's left is to enjoy your beautiful hardwood floor.

To ensure that your floor will give both beauty and function for many years, you'll want to maintain it properly. That's the subject of Chapter 5.

Chapter **5**

Maintaining hardwood floors

*Y*ou've put money and effort into your beautiful new hardwood floor. To get the greatest return on your investment, you must take care of the floor. It is not difficult to do so. In fact, you'll find that maintaining hardwood floors is easier than taking care of most other types of flooring materials.

BASIC HARDWOOD FLOOR CARE

There are a number of things you can do without touching a mop or broom that will give longer life to your hardwood floor.

Humidity and ventilation are critical considerations for your new wood floor. Relative humidity of 40 to 50% is normally required for a long, trouble-free life. If the humidity rises over 50%, prompt air circulation should be initiated by opening interior doors and windows and by activating the ventilation system. Don't draw warm, moist air from outdoors, however, because excessive humidity will cause the wood to expand.

The summer months are especially critical. Inspect your wood floors regularly. If necessary, turn on the heating system. If less than a 35% humidity level persists, use humidification to prevent excessive dryness and possible wood shrinkage.

When excessive tightening of the floor becomes noticeable, call your flooring contractor or supplier immediately. When unusually wide cracks begin to appear, you should also call your flooring contractor or supplier to solve the problem as quickly as possible.

Be sure that your air-conditioning system is operating within the 40 to 50% range of normal relative humidity. Ventilation equipment should be available for year-round use.

Avoid exposing the wood floor to water during periods of inclement weather by protecting your floors at exterior doorways. Floor protection should be checked thoroughly to ensure that no moisture is trapped underneath. Windows and doors should be closed during rainy weather. All leaks must be corrected immediately.

DAILY FLOOR CARE

Wood floors, when properly finished, are the easiest of all floor surfaces to keep clean. And, unlike carpeted or resilient floors that show age regardless of care, wood floors can be kept looking like new, year after year, with minimum care.

Both open-grained and close-grained woods are used in flooring. Heading the list of hard, open-grained woods is durable oak, which is used for an estimated 95% of all wood floors. Other hardwoods include northern walnut, pecan, ash, elm, and chestnut. Among the close-grained woods are maple, birch, beech, Douglas fir, and yellow pine. Since the overwhelming majority of wood floors are hardwood, I'll cover the daily care specifically required by this type of flooring.

A good rule of thumb for minimum care is to vacuum or dust-mop weekly. An occasional buffing helps to remove scuff marks that may appear in the wax coating. Rewax once or twice a year, or as often as needed in high-traffic areas, with a liquid buffing wax/cleaner combination.

No matter what finish your wood floor has, or what claims the manufacturer makes for the finish, never wash or wet-mop wood floors. Water can seep between the boards and leave dark stains. Sometimes it also can warp the boards.

Always read the label of the products you use. The recommendations made here on care serve as general guidelines for selecting and using floor maintenance materials. Except for any directions on using water on wood, wherever these guidelines vary from the product's label instructions, always follow the label.

UNDERSTANDING FLOOR FINISHES

As you learned in the last chapter, there are two principal types of finishes that are used on wood floors: penetrating sealers and surface finishes. Each requires about the same care, but when it comes to removing stains or restoring the finish in heavy-traffic areas, the methods vary. It helps, therefore, to know the type of finish that was used on your floors. If you just put the finish on, as described in Chapter 4, you already know. Maybe you purchased a home with an existing hardwood floor, however. How can you find out what type of floor finish you have?

As a general rule, you can be sure your floor was prefinished at the factory if it has V-shaped grooves along the edges where the boards join and sometimes where the ends butt. Unfinished plank flooring may also have grooved edges, but usually this is a mark of a prefinished floor. Once you have established that the floor was prefinished, the flooring manufacturer can tell you whether or not it has a penetrating, sealer finish.

If the floor has no grooves, it was in all likelihood finished after it was installed by a local craftsmen or a do-it-yourselfer. To determine what kind of finish was used, call the builder, floor finisher, or previous owner, if possible. When in doubt, it is safest to assume that a surface finish was used. Treating a penetrating, sealer finish as though it were a surface finish can do no harm, whereas a surface finish treated as a penetrating sealer is likely to be ground away.

MAINTAINING PENETRATING SEALERS

A penetrating sealer is the finish that is recommended for most residential floors. As its name implies, the sealer soaks into the wood pores and then hardens to seal the floor against dirt and certain stains.

At the surface, the sealer delivers a low-gloss satin finish that wears only as the wood wears. Because of this, color may be added to the liquid sealer at the time of application. The eventual effect of traffic will be far less apparent than with other finishes that only coat the surface. When an area does begin to show wear, it can be refinished easily. The new application will blend into the old without lap marks or other signs of repair.

The beauty and wear resistance of wood floor finishes with a penetrating sealer may be further enhanced by wax. A wax coating forms a barrier against the most frequent kinds of abrasion and can be renewed easily.

A penetrating sealer also may be used as an undercoat for surface finishes. It serves as a stain that colors the wood before the surface finish is applied. The surface finish used should be compatible with the penetrating sealer. An unmatched surface coating may peel.

MAINTAINING WATER-BASE AND OIL-BASE FINISHES

Surface finishes include polyurethane, varnish, shellac, and lacquer. The maintenance of these finishes is different from that of penetrating sealers.

Polyurethane is a blend of synthetic resins, plasticizers, and other film-forming ingredients that produces an extremely durable surface that is moisture-resistant. It is the best choice for a kitchen where the floor is exposed to stains and spills. Polyurethane is available in both high-gloss and matte finishes.

Some manufacturers of polyurethane products say no waxing is required. Many flooring experts, however, believe that you'll get better wear and appearance if you give it the same care as other surface finishes.

Varnish finishes may be high, medium, or low in gloss. Varnish tends to darken with age and is difficult to touch up. It dries slowly. If the quality is good, a varnish finish will provide a highly durable surface. If not, it tends to become brittle, to powder, and to show white scars.

Shellac is a popular finish for floors in houses that are built in certain areas of the country. It dries so fast that two coats can easily be applied in one day and the floor used 8 hours later. Liquid spills, however, can cause hard-to-remove spots on a shellac finish. The abrasive action of footsteps also creates frictional heat that softens the finish and permits the entry of dirt. Waxing is essential to protect the finish.

Even faster drying than shellac, lacquer requires real skill to apply. It produces a tough, high sheen, but this sheen is difficult to maintain and scuff marks shown easily.

MAINTAINING POLYMER FINISHES

There is a third classification of finishes that has been developed recently that is known as an irradiated polymer. Thus far, it is used primarily in commercial applications. Each brand of flooring that uses a polymer finish has a different maintenance schedule, which is available from the manufacturer.

EASY FLOOR CARE

If your floors are new or newly refinished with either a penetrating sealer or a surface finish, start them off right by applying a liquid buffing wax/cleaner or a coating of paste wax. The wax will form a protective barrier over the finish and keep out dirt and potential stain-causing matter, so that your floor will stay beautiful and resist wear for a long time.

Liquid buffing wax is easier to use than paste wax and so will probably be used more often. For this reason, many professionals recommend the liquid with these two cautions: the liquid or paste wax must be designated for use on hardwood floors, and the liquid wax must not have a water base. Check the label since some manufacturers recommend their water-based products for wood, while many professionals believe only a solvent-based product should be used. Solvent-based waxes will have the odor of a dry cleaning fluid.

Follow the manufacturer's directions. Apply the wax and buff it well, preferably with a 12-inch machine buffer that is available from rental companies. You may want to buff small areas by hand with clean cloths.

Vacuuming is the best way to remove surface dust and dirt before it gets walked into the wax and dulls its luster. Vacuuming also pulls accumulated dust from the grooves of prefinished and plank floors. When floor luster has dulled a bit and scuff marks begin to show, you can restore the original beauty often, without adding new wax, by simply machine or hand buffing.

After 4 to 6 months of wear, inspect your floors closely to see if there's been a dirt build-up, or if the wax has discolored. If your floors were originally finished in a dark tone, you may see a lightening of the finish in traffic areas. If none of this is apparent, just apply a new coat of wax over the old and buff it well to restore the luster. If such conditions do exist, the following procedures can be followed to correct them.

Use a combination liquid cleaner/wax. Make sure that it has a solvent rather than a water base. For dark floors, choose a buffing wax in a compatible dark color. Spread it with a cloth or fine steel wool. Rub the area gently to remove grime and the old wax, then wipe it clean. Let the solution dry 20 minutes or so and then buff. If dull spots remain after buffing, apply a second coat and repeat.

If your floors were stained, it's always a good idea to use a colored wax or cleaner to help maintain the original color. Check the floor-care products in your local stores.

MAINTAINING SPECIAL SURFACES

Distressed wood floors have been wire brushed to remove the soft portion of the wood and give it an antique, textured appearance. The resulting uneven surface tends to trap dirt. Recommended care is a regular sweeping with a stiff broom. Follow by vacuuming to pick up the loosened dirt.

Such floors are usually stained a dark color to further convey the aged wood effect. What remains after the wire brush treatment, however, are only the toughest wood fibers. These are somewhat resistant to penetration by the finish color, which means a need for more frequent color renewal. This can be accomplished by using a wax or cleaner/wax combination of the proper color to maintain the original color tone.

REMOVING STAINS

Most stains can be prevented or minimized by keeping the floors well waxed and by wiping up any spilled liquid immediately. Here are some first-aid suggestions for common accidents. When removing a stain, always begin at the outer edge and work toward the middle to prevent it from spreading.

Dried milk or food   Rub the spot with a damp cloth. Then rub dry and rewax.

Standing water   Rub the spot with No. 00 steel wool and rewax. If this fails, lightly sand the area with a fine grit sandpaper. Clean the spot and the surrounding area using No. 1 steel wool and mineral spirits or a proprietary floor cleaner. Let the floor dry. Apply a matching finish to the floor and feather it out and into the surrounding area. Wax after the finish dries thoroughly.

Dark spots   Clean the spot and surrounding area with No. 1 steel wool and a good floor cleaner or mineral spirits. Thoroughly wash the spotted area with household vinegar. Allow the vinegar to remain for 3 to 4 minutes. If the spot remains, sand it with a fine grit sandpaper, feathering out 3 to 4 inches beyond the stain into the surrounding area, then rewax and polish.

If repeated applications of vinegar do not remove the spot, apply an oxalic acid solution directly onto the spot. Proportions are 1 ounce oxalic acid to 1 quart water, or fractions thereof.

Caution   Oxalic acid is a poison. Use rubber gloves. Pour a small amount directly on the spot and let the solution stand 1 hour. Sponge the spot with clear water. A second treatment may be helpful if the spot refuses to yield.

If a second application of oxalic acid fails, sand the area with No. 00 sandpaper and apply a matching finish. Feather the finish out and into the surrounding floor area. Let dry. Buff lightly with No. 00 steel wool. Apply a second coat of finish, let it dry, and then wax. If the spot is still visible, the only remaining remedy is to replace the affected flooring.

Note   Oxalic acid is a bleaching agent. Whenever it is used, the treated floor area will probably have to be stained and refinished to match the original color.

Heel marks, caster marks, etc.   Rub vigorously with fine steel wool and a good floor cleaner. Wipe dry and polish.

Ink   Follow the same procedures as for other dark spots.

Animal and diaper   Spots that are not too old may sometimes be removed in the same manner as other dark spots. If spots resist cleaning efforts, the affected flooring can be refinished.

Mold   Mold or mildew is a surface condition that is caused by damp, stagnant air. After seeing that proper ventilation is provided for the room, the mold can usually be removed with a good cleaning fluid.

Chewing gum, crayon, candle wax   Apply ice to the stain until the deposit is brittle enough to crumble. A cleaning fluid poured around the area, but not on it, soaks under the deposit and loosens it.

Cigarette burns   If they are not too deep, steel wool will often remove them. Moisten the steel wool with soap and water to increase its effectiveness.

Alcohol   Rub with a liquid or paste wax, silver polish, boiled linseed oil, or a cloth that has been dampened barely in ammonia. Rewax the affected area.

Oil and grease   Rub on a kitchen soap that has a high lye content, or saturate cotton with hydrogen peroxide and place it over the stain. Saturate a second layer of cotton with ammonia and place it over the first. Repeat until the stain is removed.

Wax build-up   Floors that have not had proper care may acquire a wax build-up. Strip all the old wax away with mineral spirits or naptha. Use cloths and a fine steel wool to remove all residue before you apply a new wax. It's a good idea to perform this complete stripping job every now and then instead of using the liquid cleaner/wax process. Stripping removes all the old wax and dirt, which builds up inevitably over a period of time and partially hides the beauty and color of the wood grain.

Caution   Naptha is extremely flammable. Use it only where there is no open flame or danger of spark and provide ample ventilation.

REPAIRING FLOOR FINISHES

Small areas of floors finished with a penetrating sealer can be required successfully without professional help. With special care and skill, you may also be able to repair varnish and polyurethane finishes yourself. Such repair may be necessary after a stain has been removed or there has been water damage. Use steel wool to smooth out the affected boards and feather an inch or two into the surrounding area. Then brush on one or more thin coats of finish. Feather it into the old finish to prevent lap marks. Allow plenty of drying time between coats and then wax well.

Caution Don't attempt this if you have a lacquer or shellac finish since these are almost impossible to patch successfully. For a small, relatively inconspicuous area, you might get by with a steel wool cleaning that is followed by a paste wax. You won't get an exact match, but it could serve as temporary repair. The alternate is sanding to expose bare wood over the entire room and applying a new finish. Such refinishes will be covered in detail in Chapter 6.

REPAIRING CRACKS AND SQUEAKS

All the wood in your home will contract or expand according to the moisture in the air. Doors and windows may swell and stick during rainy seasons. In dry, cold weather, cracks and fine lines or separations may appear in wall cabinets and furniture. This is characteristic of wood because wood is a product of nature and its natural quality is what makes it desirable.

The same reaction to humidity or the lack of it is happening constantly in your wood floors. Tiny cracks between the edges of the boards may appear when unusually dry conditions are produced by your heating system. This condition can usually be corrected simply by installing a

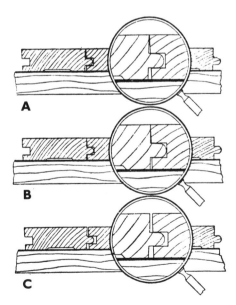

5-1 How moisture content affects expansion and contraction, and thus cracks, in hardwood flooring.

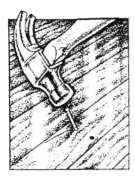

5-2 Drive finishing nail through pilot holes to nail down a squeak (courtesy National Oak Floor Manufacturers Association).

humidifier. With the proper balance of moisture content in the house, both family and floors will benefit from a healthier environment (See FIG. 5-1).

When interiors become damp in rainy weather, the boards may expand so that the edges rub together and produce a squeak. An improperly fastened floor or subfloor can also cause squeaks.

To correct this situation, first try lubrication. Put a liberal amount of powdered soap stone, talcum powder, or powdered graphite between adjacent boards where the noise occurs. Another method is to drive triangular glazier points between the strips and to use a putty knife to set them below the surface.

If that method doesn't work, drive 2-inch finishing nails through pilot holes that have been drilled into the face of the flooring (FIG. 5-2). Nails should go through both edges of the boards. Set them with a nail punch and hide them with matching color putty and wax.

If these efforts don't solve the problems of squeaks or cracks, repairs must be made. Let's consider how the repairs for these and other hardwood floor problems can be accomplished by the do-it-yourselfer.

COMMON FLOOR REPAIRS

As you have learned, the floors in most homes are composed of two separate layers. The bottom layer is called the subfloor. It might be made of rough tongue-and-groove lumber that is nailed directly to the floor joists, while in others it will run at right angles to the joists. A layer of building paper covers the subfloor to keep out dust and dirt. The finish or hardwood floor is laid over this paper. The finish floor runs at right angles to the subfloor and is nailed to it. The finish floor can be of either planks, blocks, or strips of hardwood.

REPAIRING CREAKING FLOORS

In most cases, a creaking floor is caused by the loosening of the nails that hold the subfloor to the joists. The nails may either pull loose or be loosened by the shrinkage of the wood. Creaking is usually in the subfloor, but will sometimes occur in the finish floor, particularly if the floor was put down before the wood was completely seasoned.

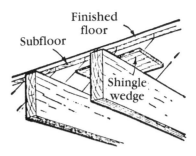

5-3 Drive a small wedge between the joist and loose board to stop a creaking floor.

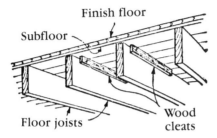

5-4 Using wood cleats to stop squeaking floors.

If the creak is in the subfloor and the underside is exposed, as when the flooring functions as the ceiling for an unfinished basement, drive a small wedge between the joist and the loose board (FIG. 5-3). The wedge will take up the play in the board and the noise will stop. If several boards are loose, nail a piece of wood to the joist high enough to prevent these boards from moving down (FIG. 5-4). The nail heads will keep the boards from moving up and effectively end the noise.

In many cases, it is impossible to reach the subfloor without tearing up the finish floor or moving a ceiling. Since neither of these alternatives is feasible, the only alternative is to try to locate the floor joist by tapping on the floor. If a joist near the creak can be located, then 2- or 3-inch finishing nails can be driven through the finish floor and subfloor and into the joist (FIG. 5-5). Drive the nail at an angle and, when it is near the surface of the floor, use a nail set to drive it below the surface of the wood so that you don't hit the finish floor with a hammer and mar the finish. Make the nail inconspicuous by filling the hole with putty or stain that matches the rest of the floor.

Occasionally, a creaking floor will be caused by a loose board in the finish floor. This board can be located by its movement when weight is

5-5 Stop squeak by driving finish nails into joists.

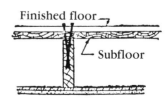

placed upon it. Use 2-inch finishing nails and drive them in at an angle. Then use the nail set in the manner just described.

Sometimes, boards in the finish floor warp to such an extent that they pull away from the subfloor and bulge. They can be driven back into place by putting a piece of heavy paper and a block of wood over them and tapping the wood sharply with a hammer. The piece of wood prevents the hammer from damaging the floor. Take care when doing this, however, for the thin edges of tongue-and-groove boards easily can be split.

REPAIRING SAGGING FLOORS

When a sagging floor is found in a very old house, it is generally because the floor joists and girders have been weakened by rot or insects. In a new house, a weak floor can, in most cases, be blamed on the builder. A floor built on undersized materials and tacked together will be neither substantial nor capable of bearing much weight.

In dealing with a weak and sagging floor, you will first have to raise it to its proper level. If it is the first floor and there is a basement underneath it, then the work is in the range of the do-it-yourselfer. Use heavy lumber and a screw jack to accomplish the work (FIG. 5-6). The size of the lumber should be about 4×4 inches. Place one of the 4×4 timbers on the basement floor, directly under the sag, and put the screw jack on top of it. This beam will distribute the weight of the floor over a relatively large

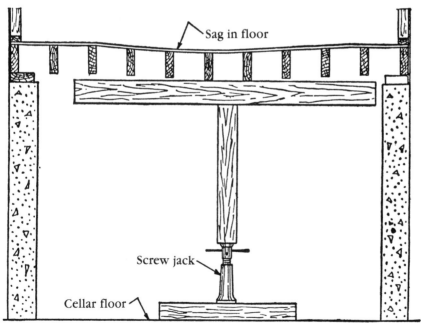

5-6 Lifting sagging floor with a screw jack and 4×4 wood members.

portion of the basement floor. If the basement floor is of heavy concrete, this step will not be required. Nail a piece of 4×4 along the sagging joists. Use a third piece of timber as a vertical beam from the top of the jack to the under portion of the 4×4 nailed to the joists. Turn up the jack until the floor is level.

Don't attempt to bring the floor to a level position all at once. If you do so, you are almost sure to crack the plaster walls and ceiling in the room above. If you raise the jack only a fraction each week, you will avoid doing extensive damage to the room above.

Check the position of the floor with a level. When it is correct, measure the distance from the bottom of the horizontal 4×4 to the floor of the basement. Cut a piece of 4×4 to this length. Turn the jack up enough to allow this beam to stand on end under the horizontal 4×4. Make sure that it is perfectly vertical and that it rests firmly on the floor. Remove the jack, along with the other timbers, and leave only the one vertical and one horizontal 4×4.

If one entire floor is sagging, it will probably be necessary to use more than one vertical support. In this case, place a vertical 4×4 under each end of the horizontal beam.

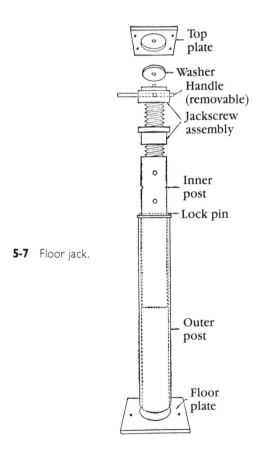

5-7 Floor jack.

Another means of raising a floor is to use metal posts that have screw jacks built into them (FIG. 5-7). The post is provided with two plates, one of which rests on the basement floor, while the other fits between the top of the post and the joist or girder that is to be raised. These posts are made so that they can be adjusted to different heights. Once the floor has been brought to the right level, the posts can be left as a permanent support. As before, turn the jack only a small amount each week so that the floor will be raised slowly.

When part of the total weight of a floor and the objects on it is supported by posts, it is important that each post have the proper footing. Most concrete basement floors are rather thin, so it is often necessary to prepare the floor before you install the posts.

To make a substantial footing for the posts in the basement floor, break up about 2 square feet of the concrete floor at the point where the post is to stand. Do this work with a heavy hammer or a piece of pipe. Once the surface is broken, dig a hole about 12 inches deep and fill it with concrete that is made with 1 part cement, 2 parts sand, and 3 parts coarse aggregate. Level this with the floor and make a smooth surface. Allow about 1 week for the footing to dry before you place the posts on it. Cover the concrete during this period and keep it moist.

Fortunately, most defects are associated with the first floor, so the basement or crawl space underneath it allows for the installation of posts and other kinds of reinforcement. Sagging floors above the first floor level cannot be practically remedied, short of taking up the flooring and making extensive repairs. For this situation, it's best to call in a good carpenter to do the work.

REPAIRING FLOOR CRACKS

If the wood in a matched hardwood floor is properly seasoned, very few cracks should appear between the boards. Many houses, however, are equipped with plank floors in which cracks of varying size are almost sure to appear between each board. In very old houses, these cracks can be quite large. There are several kinds of plastic fillers, but many of them tend to shrink and crack as they dry. A good filler can be made of sawdust and wood glue that has been mixed into a paste. If possible, the sawdust should be of the same wood as the flooring.

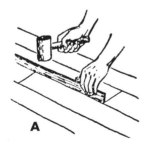 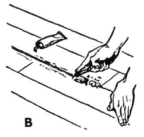

5-8   Filling flooring cracks with a wood strip (A) and plastic crack filler (B).

Before you attempt to fill a crack, clean it out. Any dirt in it will prevent the filler from adhering to the wood. Pack the filler in tightly so that it stands slightly higher than the surface of the floor. After it is dry, sand the top to the floor level and apply a little stain to match the finish to the rest of the floor. Over very wide cracks, glue a thin strip of wood and sand or plane it to match the floor surface (FIG. 5-8).

REPLACING FLOORING BOARDS

Wood floors become uneven from excessive wear. This can leave high places, particularly where knots and the heads of nails occur, since these possess a greater resistance to wear. An uneven surface occasionally occurs along the edges of the floor boards, which become raised due to the boards curling up as they warp. If the underlying joists warp and twist, or if there is some settlement of the foundation walls on which these joists rest, the level of the floor may be redistributed.

To replace boards, first study the diagram of a single wood plank floor that is given in FIG. 5-9. The joists are usually about 2 inches wide. The boards, which cross the joists at right angles, are nailed at the centerline of the joist. If the ends of two boards meet in the form of a heading joint on one joist, both boards must be nailed to the joist.

To remove the boards without damage is not easy. If one board is to be discarded, you can bore a round hole as near as convenient to the side of the joist and use a keyhole saw and compass saw to cut one board close up to the joists. If the other end of the board runs to a heading joint on another joist, work back along the board. Pry it up at the intervening joist and take it off the joist where it ends.

Perhaps only part of the lifted board is defective, in which case you can cut it across to end on a suitable joist, leaving it ready for replacement later. When prying up the board, the nails will most likely be pulled up out of the joist. Rest the board, bottom side up, on a stool or sawhorse

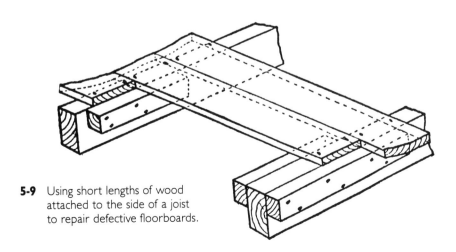

5-9 Using short lengths of wood attached to the side of a joist to repair defective floorboards.

and tap the nails back sufficiently for the heads to be gripped by pincers or by the claws of the hammer. Obstinate boards may have to have the nails punched fully into the joist to free the board.

If a heading joint is not conveniently near, the board may have to be cut through at two places in order to remove the defective part. Bore a 1/2-inch hole and cut across the board with a keyhole saw.

Find the run of the joists, indicated by the position of the nail heads. On the assumption that they mark approximately the middle of the joist, measure 1 inch to either side of the nail, then square a line across with a square. Mark it with a pencil. Put a fine brad awl through the board about 1/4 inch away from the pencil line, on the free side, then bore a hole. If these dimensions have been correctly established, the joist will be visible and the saw can be put through the cut alongside the joist and across the board. As soon as the cut is long enough, take out the keyhole saw and insert a compass saw or a small crosscut saw and complete the cut.

Beware of water pipes, gas lines, and electric wiring when you cut the boards. They usually run in the space between the joists. When there is a room below the floor where the work is in progress, some guide to the position of pipes and cables can be estimated from the location of lighting and plumbing fixtures in that room.

Having cut the board, the ends of the fixed parts can be trimmed with a sharp chisel to a square edge. If several boards have to be cut away, take them back to the joists that are one or two away to the right or left of the one originally selected for the patch. In other words, break the joints so that a board that extends over a given joist is next to one at which a board ends, and so on. Thus you will not get a weak line of joints running along the same joist.

A typical job is shown in FIG. 5-10. The joists are numbered and the floor boards lettered. A heading joint is shown at XX on board B. It is not always practical to make heading joints when you replace the boards. Often, the best thing to do is to support the ends of the replacement boards by nailing or screwing a strong cleat to the side of the joist where the end of the new board will rest. The cleat should be at least 1 1/2 inches thick and about 3 inches wide. Take it halfway along and under the

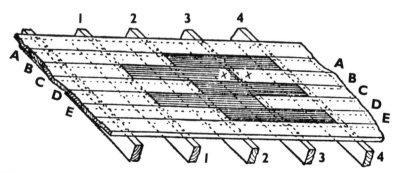

5-10 Replacing defective floorboards between joists 2 and 3 requires the removing and replacing of all shaded boards in this typical example.

boards that adjoin the one that it will support.

Two boards can be cut through obliquely when they have to be jointed over a single joist. In this instance, it is assumed that both boards can be taken to the bench and cut by a tenon saw to a suitable angle. This makes a neat and sound job, with the nails being driven through the oblique portion. The angle can be marked across the edges of the board with an adjustable level.

REPAIRING BLOCK AND PARQUET FLOORS

Loose blocks in a floor, if there are not many, should be removed. It will then be possible to scrape off the old mastic underneath. Put in fresh mastic, which you can get at hardware stores, and bed in the block. If the defect is extensive, the repairs may be more than an amateur can undertake successfully.

Parquet floors are glued and then usually screwed into place. Dampness may cause parts of the design to come loose. In such cases, the cause of the dampness should be found before you attempt a remedy. Wood glue can be used to hold the different parts of a pattern together if a whole unit is defective. These diamonds and other designs are bedded upon a piece of low-quality material that has an open weave that helps to hold them together. It will probably be best to unite the various parts of an element first, and then let the glue harden before relaying it to ensure that all joints are firmly set and safe to handle.

Those are the basics of maintaining and repairing new and old hardwood floors. You can see that while repairs can be made by the do-it-yourselfer, they are not easy. Therefore, the best insurance against major flooring repair is good maintenance methods performed on efficiently selected flooring.

You may someday need to refinish your new or old hardwood floor, however. Complete instructions on how to do so are presented in Chapter 6.

Chapter **6**

Refinishing hardwood floors

*T*here has been a revival of interest in older homes. Many people are spending their weekends replacing molding, stripping floors, insulating walls, refinishing woodwork, and revitalizing the works of past craftsmen. Others are remodeling older homes for economy and function and finding that the old ways are often the best.

In this final chapter, you'll learn how to refinish a hardwood floor. In most cases, it can be done easily by the average do-it-yourselfer with an understanding of the process and some rented equipment. Take your time. Learn how it's done. Talk with your supplier and to others who have refinished hardwood floors. You'll find a wealth of information and opinions on how best to renew hardwood flooring.

If you haven't already, read Chapter 4 on how to finish hardwood floors. It discusses the types of finishes that are commonly used, the equipment needed, how the finishes are applied, and other background information that is vital to the flooring restorer. It also includes numerous photos of how an old hardwood floor was restored.

WHEN TO REFINISH

Where floors have become badly discolored and worn by neglect or improper maintenance, the most practical procedure, and often the only one that will restore a fine finish, is to have the old finish removed and the floor reconditioned by power sanding (FIG. 6-1). Where the floors have been reasonably well-maintained but the finish has become dingy with age, refinishing without power sanding may be practical. The method for removing the old finish depends upon the kind of finish that was originally used.

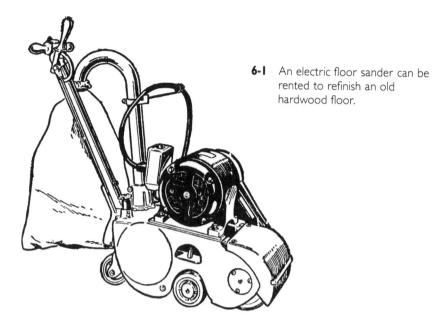

6-1 An electric floor sander can be rented to refinish an old hardwood floor.

VARNISH FINISHES

Old, discolored varnish is usually removed most easily by power sanding. If desired, this can be done with a liquid varnish remover. Alkaline solutions in water and removers sold in powder form that are dissolved in water should not be used.

Follow the directions for the liquid remover carefully. Since some of the old, discolored varnish remains embedded in the wood, do not expect a complete restoration of the natural wood. Clean traffic channels where the old varnish has worn through and where dirt has been ground into the wood thoroughly by sanding.

SHELLAC FINISHES

Old shellac and wax finishes that have merely become soiled by dirt that has lodged in the wax coating may be cleaned by going over the floor with steel wool that has been saturated with clean turpentine. Any white spots in the shellac that have been caused by contact with water may be taken out by rubbing the spot lightly with a soft cloth that has been moistened with denatured alcohol. The alcohol must be used with care, however, to avoid cutting the shellac coating.

On floors where the dirt is ground into the shellac itself or where white spots have penetrated the coating, more drastic treatment is necessary. First, wash the floor with a neutral or mildly alkaline soap solution. Then scour the floor with No. 3 steel wool and denatured alcohol. If the floor boards are level and are not warped or cupped, the scouring can be done to advantage with a floor polishing machine that is fitted with a wire brush to which a pad of the No. 3 steel wool is attached. After scouring, the floor should be wiped clean and allowed to dry thoroughly before you refinish it with shellac or another finish.

OIL FINISHES

Some older floors have been finished with linseed oil. To refinish, clean the surface and apply a new coating of oil. Make needed repairs and smooth rough spots if the floor is in poor condition.

To remove oil, some professionals recommend that you wet about 10 square feet of floor with a mop and warm water and then liberally sprinkle the area with a mixture of one part soap powder and three parts trisodium phosphate. Scrub the floor with a stiff brush. Use only as much water as needed to form an emulsion, which will float the oil to the surface. As the oil is loosened, remove it with a squeegee and mop. Rinse the area and mop it dry. Treat the other sections in the same way. When the entire floor has been cleaned, let the surface dry at least 24 hours and then sand it with a machine and finish it as desired.

A reminder Many professional floor refinishers recommend that no water touch the wood since it is quickly absorbed and may cause the wood to crack and swell once the finish is installed. Discuss this with your floor supplier.

REFINISHING PROCEDURES

Those are some of the ways that you can refinish your hardwood floor with little or no need for a floor sander. In many cases, however, it is best to roll up your sleeves and plan on renting a floor sander for a day or two in order to completely strip an old floor. TABLE 6-1 suggests the selection of sandpaper you should buy when refinishing an old floor.

First, remove all the furniture, rugs, and draperies from the room. If you're planning on painting or applying a wallcovering, do that work before you refinish the floor so the paint or paste won't drip on the new floor. Vacuum the floor.

Look for protruding nail heads and drive them down with a nail set. Tighten any loose boards by face-nailing them, preferably into joists, and then countersinking them with a nail set.

Table 6-1 Sanding Old Floors

Floor	Operation	Type of Paper	
Covered with varnish, shellac, paint, etc.	First Cut	Coarse	$3^{1}/_{2}$ (20)
	Second Cut	Medium	$1^{1}/_{2}$ (40)
	Finish Sanding	Fine	2/0 (100)

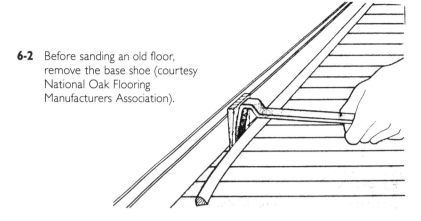

6-2 Before sanding an old floor, remove the base shoe (courtesy National Oak Flooring Manufacturers Association).

Before sanding an old floor, remove the base shoe (FIG. 6-2). Use a wood wedge behind the pry bar to protect the baseboard from damage. Figure 6-3 illustrates the cross section of a typical hardwood block floor.

SANDING AN OLD FLOOR

Most oak and many other types of flooring can be sanded and refinished a number of times. Thinner floors should be refinished with caution because repeated sanding could wear through to the subfloor. To deter-

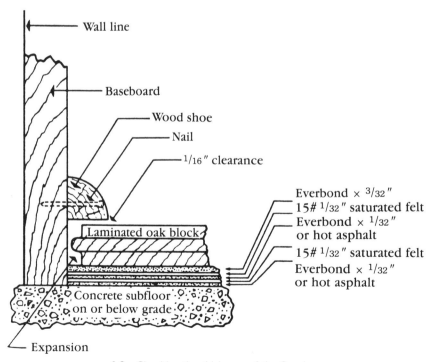

6-3 Checking the thickness of the flooring.

mine the floor thickness, remove a floor heating register or the shoe mold and baseboard (FIG. 6-3) so that an edge of the flooring is exposed and can be measured.

When you are refinishing thinner floors—$1/2$ or $3/8$ inch—use the floor polishing machine and screen abrasive rather than the drum sander. Remove as little of the surface as possible. Do not do this for face-nailed $5/16$-inch square-edged, or $5/16$-inch square-edged flooring that has been installed in mastic, however. The $5/16$-inch type is almost always face-nailed; others are blind-nailed through the tongue. The nails must be set, or driven deeper into the wood, to permit sanding of the floor.

The following instructions apply to standard $3/4$-inch strip, plank, and block floors and, with the cautions just mentioned, to the thinner materials. Use an open-faced paper to remove the finish. The heat and abrasion of the sanding operation make the old finish gummy and will quickly clog normal sandpaper. When you get down to the new wood, you can switch to a regular paper for the final finishing cuts.

The number of cuts that are required for the refinishing operation will be determined by the condition of the old floor and the thickness of the finish being removed. If the surface is in good shape and has no thick build-up of old finish and wax, one pass with the disc sander and an extra-fine sandpaper may be sufficient. Just be sure you've removed all the old finish.

If the floor is badly scarred or the boards are cupped or dished, use as many cuts as necessary to get a smooth, unblemished surface. Make the first one or two cuts at a 45-degree angle with a medium-grit sandpaper. Follow the instructions given in Chapter 4 for sanding a new floor on the succeeding cuts.

NEW FINISHES

As you learned in Chapter 4, there are two basic finishes that can be applied to hardwood floors: penetrating sealers and surface finishes. Most manufacturers of penetrating sealer finishes also make a renovator or reconditioning product you can use when traffic or other conditions cause a discoloration or wear of the finish. These products restore the floor to its original appearance without the need for sanding. They can be used on most prefinished flooring, which may be identified by the beveled edge on each strip or block.

Another way to determine if a floor was originally finished with a penetrating sealer is to scratch the surface in a corner or some other inconspicuous space with a coin or other sharp-edged object. If the finish does not flake off, a penetrating sealer was probably used and a renovator product can be applied to restore it to its original beauty.

If the condition of the flooring is very bad, sanding and the application of a complete new finish may be needed. Once the floor is completely stripped and cleaned, follow the directions in Chapter 4 for finishing hardwood floors.

With most surface finishes, the recommended method of restoration is to sand off the old finish and apply a completely new finish of a penetrating sealer or polyurethane. If the floor was originally finished with polyurethane and is in good condition, it should be cleaned and screen-disced to rough up the old finish.

Caution The adhesion of polyurethanes, epoxies, and ureaformalde-hyde finishes is affected by wax and grease, as well as some types of stains, bleaches, or sealers. Further more, one type of polyurethane may not be compatible with another type. TABLE 6-2 offers information on the grain of various types of wood and notes on finishes that may guide you in refinishing your old hardwood floor.

PLANING AN OLD FLOOR

Sometimes you will run across a strip or block of hardwood flooring that needs to be planed before it can be refinished. Let's consider how this is done.

The plane is the most extensively used hand shaving tool. The large family of planes includes bench planes and block planes, which are

Table 6-2 Wood and Wood Finishes

| | Type . . . . | Soft | Hard | | |
	Grain	Closed	Open	Closed	Notes on Finishing
Name of Wood					
Ash		....	X	....	Requires filler.
Alder		X	....	....	Stains well.
Aspen		....	....	X	Paints well.
Basswood		....	....	X	Paints well.
Beech		....	....	X	Paints poorly; varnishes well.
Birch		....	....	X	Paints and varnishes well.
Cedar		X	....	....	Paints and varnishes well.
Cherry		....	....	X	Varnishes well.
Chestnut		....	X	....	Requires filler; paints poorly.
Cottonwood		....	....	X	Paints well.
Cypress		....	....	X	Paints and varnishes well.
Elm		....	X	....	Requires filler; paints poorly.
Fir		X	....	....	Paints poorly.
Gum		....	....	X	Varnishes well.
Hemlock		X	....	....	Paints fairly well.
Hickory		....	X	....	Requires filler.
Mahogany		....	X	....	Requires filler.
Maple		....	....	X	Varnishes well.
Oak		....	X	....	Requires filler.
Pine		X	....	....	Variable depending on grain.
Teak		....	X	....	Requires filler.
Walnut		....	X	....	Requires filler.
Redwood		X	....	....	Paints well.

Note: Any type finish may be applied unless otherwise specified.

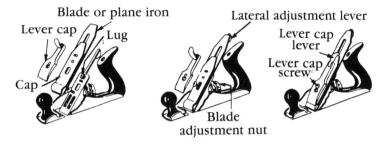

6-4 Components of a typical bench plane.

designed for general surface smoothing and squaring, and other planes that are designed for special types of surface work.

The principal parts of a bench plane and the manner in which it is assembled are shown in FIG. 6-4. The part at the rear that you grasp to push the plane ahead is called the handle. The part at the front that you grasp to guide the plane along its course is called the knob. The main body of the plane, which consists of the bottom, the sides, and the sloping part that carries the plane iron, is called the frame. The bottom of the frame is called the sole and the opening in the sole, through which the blade emerges, is called the mouth. The front end of the sole is called the toe, and the rear end is the heel.

There are three types of bench planes: the jointer plane or fore plane (FIG. 6-5), the jack plane (FIG. 6-6), and the smooth plane (FIG. 6-7). All are

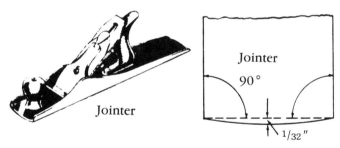

6-5 Jointer or fore plane.

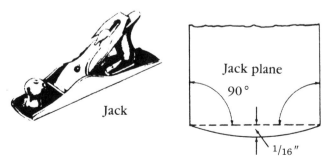

6-6 Jack plane.

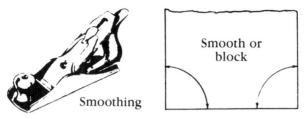

6-7 Smoothing plane.

used primarily for shaving and smoothing with the grain; the chief difference is the length of the sole. The sole of the smooth plane is about 9 inches long; the sole of the jack plane about 14 inches long; the sole of the jointer plane from 20 to 24 inches long. The longer the sole of the plane, the more uniformly flat and true the planed surface will be. Consequently, the bench plane you should use depends on the requirements with regard to surface trueness.

The smooth plane is, in general, a smoother only. It will plane a smooth, but not an especially true, surface in a short time. It is also used for cross-grain smoothing and the squaring of end stock. It is especially useful for planing smaller wood strip flooring.

The jack plane is the jack-of-all-work of the bench plane group. It can take a deeper cut and plane a truer surface than the smooth plane. It is most often used to plane larger strip or block wood flooring, that is either on or off the floor.

The jointer plane is used when the planed surface must meet the highest requirements with regard to trueness. A jointer plane would be used on the floor-level adjoining hardwood strips or blocks.

A block plane and the names of its parts are shown in FIG. 6-8. Note that the plane iron in a block plane doesn't have a plane iron cap. Also, unlike the iron in a bench plane, the iron in a block plane goes in bevel-up. The block plane, which is usually held at an angle to the work, is used chiefly for the cross-grain squaring of end stock. It is also useful for smoothing all planed surfaces on very small work.

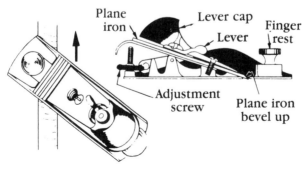

6-8 Block plane.

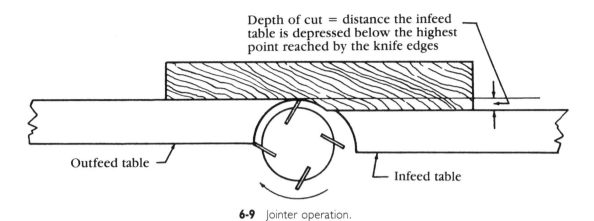

Depth of cut = distance the infeed
table is depressed below the highest
point reached by the knife edges

Outfeed table

Infeed table

6-9 Jointer operation.

Another shaving tool that you may be able to use to smooth a hardwood flooring strip or block is the jointer (FIG. 6-9). The jointer is a machine for power-planing the faces, edges, and ends of stock. The planing is done by a revolving cutterhead that is equipped with two or more knives. The table consists of two parts on either side of the cutterhead. The stock is started on the infeed table and fed past the cutterhead and onto the outfeed table. The surface of the outfeed table must be exactly level with the highest point reached by the knife edges. The surface of the infeed table is depressed below the surface of the outfeed table by an amount equal to the desired depth of the cut. A jointer could be used to shave the surface off new hardwood boards that replace old ones or to lower the depth of all hardwood floor boards in a room prior to installation.

CHISELING

As you refinish a hardwood floor or remodel to accommodate new heating vents, new walls, piping, or cabinets, you may need to change the shape of the hardwood floor that is to be repaired or installed. The common wood chisel is extremely useful for this task.

A wood chisel should always be held with the flat side, or back side, against the wood for smoothing and finishing cuts. Whenever possible, it should not be pushed straight through an opening, but should be moved laterally at the same time that it is pushed forward. This method ensures a shearing cut, which with care will produce a smooth and even surface, even when the work is across the grain. On rough work, use a hammer or mallet to drive the socket-type chisel.

On fine work, use your hand as the driving power on tang-type chisels. For rough cuts, the bevel edge of the chisel is held against the work. Whenever possible, other tools, such as saws and planes, should be used to remove as much of the waste as possible. The chisel should be used for finishing purposes only.

To chisel horizontally with the grain, grasp the chisel handle in one hand and extend your thumb towards the blade (FIG. 6-10). The cut is con-

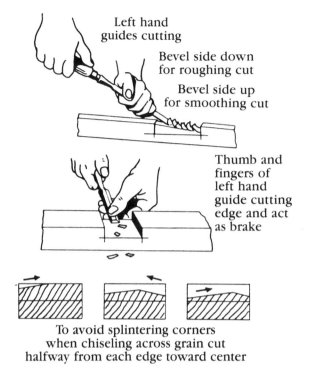

Left hand
guides cutting

Bevel side down
for roughing cut

Bevel side up
for smoothing cut

Thumb and
fingers of
left hand
guide cutting
edge and act
as brake

To avoid splintering corners
when chiseling across grain cut
halfway from each edge toward center

6-10 Chiseling horizontally with and across the grain.

trolled by holding the blade firmly with the other hand, knuckles up, and with your hand well back of the cutting edge. The hand on the chisel handle is used to force the chisel into the wood. The other hand that presses downward on the chisel blade regulates the length and depth of the cut.

To chisel horizontally across the grain, hold the work so that it does not move. Remove most of the waste wood by using the chisel with the bevel down. For light work, use hand pressure or light blows on the end of the chisel handle with the palm of your right hand. For heavy work, use a mallet. To avoid splitting at the edge of the wood, cut from each edge to the center and slightly upward so that the waste wood at the center is removed last (FIG. 6-10).

Make finishing cuts with the flat side of the chisel down. Never use a mallet when making finishing cuts, even on large work. One-hand pressure is all that is necessary to drive the chisel, which is guided by the thumb and forefinger of your other hand. Finish cuts should also be made from each edge toward the center. Do not cut all the way across from one edge to the other, or the far edge may split.

To cut a round corner on the end of a piece of hardwood flooring, first lay out the work and remove as much waste as possible with a saw (FIG. 6-11). Use the chisel with the bevel side down to make a series of straight cuts tangent to the curve. Move the chisel sideways across the work as it is moved forward. Finish the curve by paring with the level side up. Convex curves are cut in the same manner as round corners.

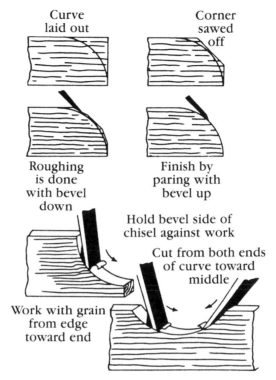

6-11   Chiseling corners and curves on hardwood flooring.

When cutting a concave curve with a chisel, remove most of the waste wood with a coping saw or a compass saw. Smooth and finish the curve by chiseling (FIG. 6-11) with the grain. Hold the chisel against the work. Press down on the chisel with the other hand and, at the same time, draw back on the handle to drive the cutting edge in a sweeping curve. Care must be used to take only light cuts, or the work may be damaged.

Vertical chiseling (FIG. 6-12) means to cut at right angles to the surface of the wood, which is horizontal. This technique is used to make a hole in a hardwood flooring strip or block to allow for the passage of a cable,

6-12   Vertical chiseling on hardwood flooring.

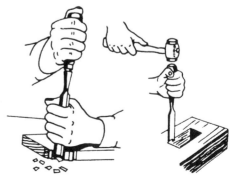

pipe, or other element. Usually it involves cutting across the wood fiber when you are chiseling vertically across the grain, using a mallet to drive the chisel.

To keep you safe when you use a chisel, here are a few reminders:

- Secure the work so that it cannot move.
- Keep both hands back of the cutting edge at all times.
- Don't start a cut on a guideline. Start slightly away from the line so that there is a small amount of material to be removed by the refinishing cuts.
- When starting a cut, always chisel away from the guideline and toward the waste wood so that no splitting will occur at the edge.
- Never cut towards yourself with a chisel.
- Make the shavings thin, especially when finishing.
- Examine the grain of the wood to see which way it runs. Cut with the grain to sever the fibers and leave the wood smooth. Cutting against the grain splits the wood and leaves it rough. This type of cut cannot be controlled.

Safe and proper chiseling and planing techniques will help you do a more professional job when you refinish your hardwood floor. That's it! You've learned to understand, plan, install, finish, maintain, and refinish hardwood floors. Now it's time to sit back and enjoy.

Hardwood floor data

TABLES A-1 through A-4 will help you fasten and measure hardwood floors.

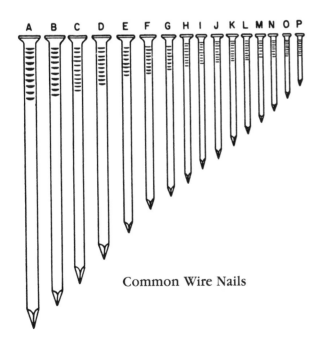

Common Wire Nails

Table A-1   Nail Sizes

Size		Length and Gage		Approximate Number to Pound
		Inches	Number	Pound
A	60d	6	2	11
B	50d	5½	3	14
C	40d	5	4	18
D	30d	4½	5	24
E	20d	4	6	31
F	16d	3½	7	49
G	12d	3¼	8	63
H	10d	3	9	69
I	9d	2¾	10¼	96
J	8d	2½	10¼	106
K	7d	2¼	11½	161
L	6d	2	11½	181
M	5d	1¾	12½	271
N	4d	1½	12½	316
O	3d	1¼	14	568
P	2d	1	15	876

Table A-2 Conversion: Feet to Meters

Feet	Meters	Feet	Meters	Feet	Meters
1	0.3048	35	10.6680	68	20.7264
2	0.6096	36	10.9728	69	21.0312
3	0.9144	37	11.2776	70	21.3360
4	1.2192	38	11.5824	71	21.6408
5	1.5240	39	11.8872	72	21.9456
6	1.8288	40	12.1920	73	22.2504
7	2.1336	41	12.4968	74	22.5552
8	2.4384	42	12.8016	75	22.8600
9	2.7432	43	13.1064	76	23.1648
10	3.0480	44	13.4112	77	23.4696
11	3.3528	45	13.7160	78	23.7744
12	3.6576	46	14.0208	79	24.0792
13	3.9624	47	14.3256	80	24.3840
14	4.2672	48	14.6304	81	24.6888
15	4.5720	49	14.9352	82	24.9936
16	4.8768	50	15.2400	83	25.2984
17	5.1816	51	15.5448	84	25.6032
18	5.4864	52	15.8496	85	25.9080
19	5.7912	53	16.1544	86	26.2128
20	6.0960	54	16.4592	87	26.5176
21	6.4008	55	16.7640	88	26.8224
22	6.7056	56	17.0688	89	27.1272
23	7.0104	57	17.3736	90	27.4320
24	7.3152	58	17.6784	91	27.7368
25	7.6200	59	17.9832	92	28.0416
26	7.9248	60	18.2880	93	28.3464
27	8.2296	61	18.5928	94	28.6512
28	8.5344	62	18.8976	95	28.9560
29	8.8392	63	19.2024	96	29.2608
30	9.1440	64	19.5072	97	29.5656
31	9.4488	65	19.8120	98	29.8704
32	9.7536	66	20.1168	99	30.1752
33	10.0584	67	20.4216	100	30.4800
34	10.3632				

Table A-3   Conversion: Millimeters to Inches

Milli-meters	Inches	Milli-meters	Inches	Milli-meters	Inches	Milli-meters	Inches
1	0.039370	26	1.023622	51	2.007874	76	2.992126
2	0.078740	27	1.062992	52	2.047244	77	3.031496
3	0.118110	28	1.102362	53	2.086614	78	3.070866
4	0.157480	29	1.141732	54	2.125984	79	3.110236
5	0.196850	30	1.181102	55	2.165354	80	3.149606
6	0.236220	31	1.220472	56	2.204724	81	3.188976
7	0.275591	32	1.259843	57	2.244094	82	3.228346
8	0.314961	33	1.299213	58	2.283465	83	3.267717
9	0.354331	34	1.338583	59	2.322835	84	3.307087
10	0.393701	35	1.377953	60	2.362205	85	3.346457
11	0.433071	36	1.417323	61	2.401575	86	3.385827
12	0.472441	37	1.456693	62	2.440945	87	3.425197
13	0.511811	38	1.496063	63	2.480315	88	3.464567
14	0.551181	39	1.535433	64	2.519685	89	3.503937
15	0.590551	40	1.574803	65	2.559055	90	3.543307
16	0.629921	41	1.614173	66	2.598425	91	3.582677
17	0.669291	42	1.653543	67	2.637795	92	3.622047
18	0.708661	43	1.692913	68	2.677165	93	3.661417
19	0.748031	44	1.732283	69	2.716535	94	3.700787
20	0.787402	45	1.771654	70	2.755906	95	3.740157
21	0.826772	46	1.811024	71	2.795276	96	3.779528
22	0.866142	47	1.850394	72	2.834646	97	3.818898
23	0.905512	48	1.889764	73	2.874016	98	3.858268
24	0.944882	49	1.929134	74	2.913386	99	3.897638
25	0.984252	50	1.968504	75	2.952756	100	3.937008

Table A-4   Fractions and
Decimal Equivalents

$1/64$	.015625	$33/64$	.515625
$1/32$	.03125	$17/32$	.53125
$3/64$	.046875	$35/64$	.546875
$1/16$	.0625	$9/16$	.5625
$5/64$	.078125	$37/64$	.578125
$3/32$	.09375	$19/32$	.59375
$7/64$	.109375	$39/64$	.609375
$1/8$	.125	$5/8$	.625
$9/64$	.140625	$41/64$	.640625
$5/32$	.15625	$21/32$	.65625
$11/64$	.171875	$43/64$	.671875
$3/16$	.1875	$11/16$	.6875
$13/64$	.203125	$45/64$	.703125
$7/32$	.21875	$23/32$	.71875
$15/64$	.234375	$47/64$	.734375
$1/4$	.25	$3/4$	.75
$17/64$	.265625	$49/64$	.765625
$9/32$	.28125	$25/32$	.78125
$19/64$	.296875	$51/64$	.796875
$5/16$	.3125	$13/16$	.8125
$21/64$	.328125	$53/64$	.828125
$11/32$	.34375	$27/32$	.84375
$23/64$	.359375	$55/64$	.859375
$3/8$	.375	$7/8$	.875
$25/64$	.390625	$57/64$	.890625
$13/32$	.40625	$29/32$	.90625
$27/64$	.421875	$59/64$	.921875
$7/16$	.4375	$15/16$	.9375
$29/64$	.453125	$61/64$	.953125
$15/32$	.46875	$31/32$	.96875
$31/64$	.484375	$63/64$	.984375
$1/2$	.5	1.0	1.0

Appendix **B**

Hardwood floor game markings

FIGURES B-1 through B-9 are specific markings that can be made on finished hardwood floors in gymnasiums and sports areas, courtesy of the Maple Flooring Manufacturers Association.

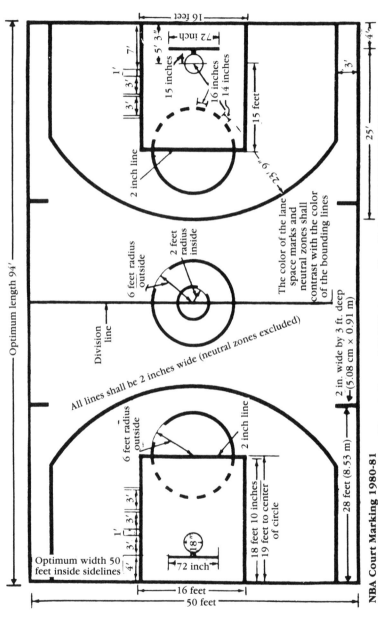

NBA Court Marking 1980-81
National Basketball Association • Olympic Tower, 645 Fifth Avenue, New York, NY 10022 (212) 826-7000

Section I—Court and Dimensions

a. The playing court shall be measured and marked in court diagram.

b. A free throw lane shall be marked at each end of the court with dimensions and markings as shown on court diagram. All boundary lines are part of the lane; lane space marks and neutral zone marks are **not**. The color of the lane space marks and neutral zones shall contrast with the color of the boundary lines. The areas identified by the lane space markings are two inches by thirty-six and the neutral zone marks are twelve inches by thirty-six inches.

c. A free throw line, 2 inches wide, shall be drawn across each of the circles indicated in court diagram. It shall be parallel to the end line and shall be 15 feet from the plane of the face of the backboard.

d. Three-point field goal area which has parallel lines three feet from the sidelines, extending from the baseline, and an arc of 23 feet-nine inches from the middle of the basket which intersects the parallel lines.

e. Four hash marks shall be drawn (two inches wide) perpendicular to the side line on each side of the court and 28 feet from the base line. The hash mark shall extend three feet onto the court.

NOTE: The National Basketball Association has issued new regulations regarding defensive play and coaches boxes. These regulations affect game line markings. Consult the NBA (212) 826-7000 for specific guidance.

B-1 NBA basketball court markings.

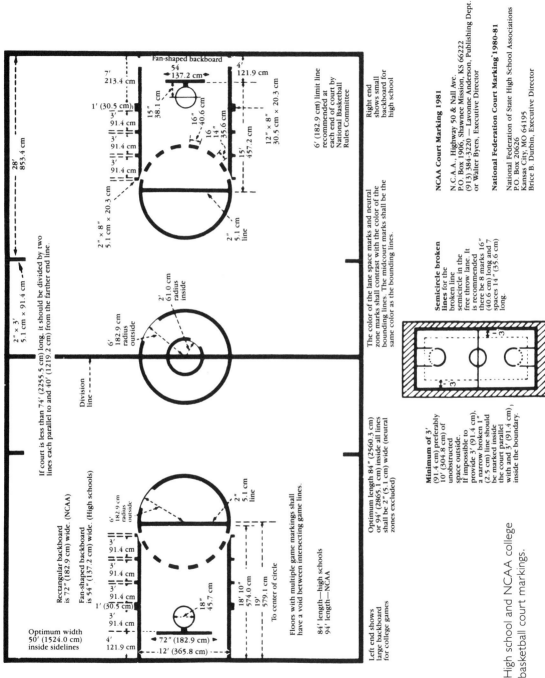

B-2 High school and NCAA college basketball court markings.

NCAA Court Marking 1981

N.C.A.A., Highway 50 & Nall Ave.
P.O. Box 1906, Shawnee Mission, KS 66222
(913) 384-3220 — Lavonne Anderson, Publishing Dept.
or Walter Byers, Executive Director

National Federation Court Marking 1980-81

National Federation of State High School Associations
P.O. Box 20626
Kansas City, MO 64195
Brice B. Durbin, Executive Director

Fan-shaped backboard

7'
213.4 cm

54
137.2 cm

4'
121.9 cm

15"
38.1 cm

1' (30.5 cm)

16"
40.6 cm

3'
91.4 cm

16"
14"
35.6 cm

3'
91.4 cm

15'
457.2 cm

3'
91.4 cm

12" × 8"
30.5 cm × 20.3 cm

6' (182.9 cm) limit line
recommended at
each end of court by
National Basketball
Rules Committee

Right end
shows small
backboard for
high school

2" × 8"
5.1 cm × 20.3 cm

2"
5.1 cm
line

28'
853.4 cm

2" × 3'
5.1 cm × 91.4 cm

If court is less than 74' (2255.5 cm) long, it should be divided by two
lines each parallel to and 40' (1219.2 cm) from the farther end line.

Division
line

2'
61.0 cm
radius
inside

6'
182.9 cm
radius
outside

The color of the lane space marks and neutral
zone marks shall contrast with the color of the
bounding lines. The midcourt marks shall be the
same color as the bounding lines.

Rectangular backboard
is 72" (182.9 cm) wide. (NCAA)

Fan-shaped backboard
is 54" (137.2 cm) wide. (High schools)

6'
182.9 cm
radius
outside

2"
5.1 cm
line

3'
91.4 cm

3'
91.4 cm

3'
91.4 cm

3'
91.4 cm

1' (30.5 cm)

18"
45.7 cm

18' 10"
574.0 cm

19'
579.1 cm

To center of circle

Floors with multiple game markings shall
have a void between intersecting game lines.

84' length—high schools
94' length—NCAA

72" (182.9 cm)

12" (365.8 cm)

Optimum width
50' (1524.0 cm)
inside sidelines

4'
121.9 cm

Optimum length 84' (2560.3 cm)
or 94' (2865.1 cm) inside all lines
shall be 2" (5.1 cm) wide (neutral
zones excluded)

Left end shows
large backboard
for college games

Minimum of 3'
(91.4 cm) preferably
10' (304.8 cm) of
unobstructed
space outside.
If impossible to
provide 3' (91.4 cm),
a narrow broken 1"
(2.5 cm) line should
be marked inside
the court parallel
with and 3' (91.4 cm)
inside the boundary.

**Semicircle broken
lines** for the
broken line
semicircle in the
free throw lane. It
is recommended
there be 8 marks 16"
(40.6 cm) long and 7
spaces 14" (35.6 cm)
long.

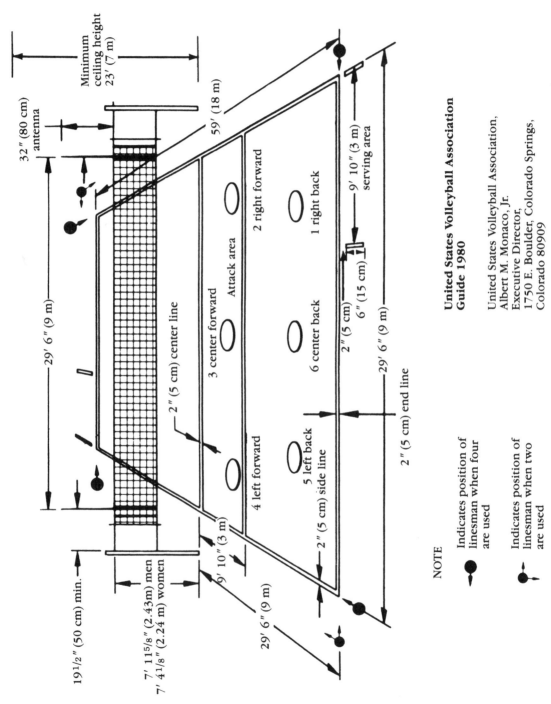

Minimum ceiling height 23' (7 m)

32" (80 cm) antenna

59' (18 m)

9' 10" (3 m) serving area

29' 6" (9 m)

2" (5 cm) center line

2 right forward

1 right back

3 center forward

Attack area

6 center back

6" (15 cm)

2" (5 cm)

29' 6" (9 m)

4 left forward

5 left back

2" (5 cm) side line

2" (5 cm) end line

19 1/2" (50 cm) min.

7' 11 5/8" (2.43m) men
7' 4 1/8" (2.24 m) women

9' 10" (3 m)

29' 6" (9 m)

United States Volleyball Association Guide 1980

United States Volleyball Association,
Albert M. Monaco, Jr.
Executive Director,
1750 E. Boulder, Colorado Springs,
Colorado 80909

NOTE

● Indicates position of linesman when four are used

↯ Indicates position of linesman when two are used

B-3   Volleyball court markings (except in high school).

HIGH SCHOOL VOLLEYBALL COURT RULES 1980-81 National Federation of State High School Associations

SECTION 2 The Court and Markings

*Art.1 .... The court shall be 60 feet long and 30 feet wide, including the outer edges of the boundary lines. An area above the court which shall be clear of any obstruction and at least 30 feet high is recommended.

Art.2 .... It is recommended all boundary lines shall be of one clearly visible color.

Art.3 .... Boundary lines shall be 2 inches wide and at least 6 feet from walls or obstacles. The end lines are the boundary lines on the short sides of the court. The sidelines are the boundary lines on the long sides of the court.

*Art.4 .... A centerline, 4 inches wide, parallel to and equidistant from the end lines, shall separate the court into two playing areas.

Art.5 .... A spiking line, 2 inches wide, shall be drawn across each playing area from sideline to sideline, the midpoint of which shall be 10 feet from the midpoint of the centerline and parallel to it.

*Art.6 .... A serving area shall be provided beyond each end line. Each shall be demarcated by two lines 6 inches long by 2 inches wide, beginning 8 inches beyond the end line and drawn perpendicular to it, one on the extension of the right sideline and the other 10 feet left of the extension. Each serving area shall be minimum of 6 feet in depth. In the event that such a space is not provided by the physical plant, the serving area shall extend into the court to whatever distance necessary to provide the minimum depth and be so marked.

*See Situations and Rulings

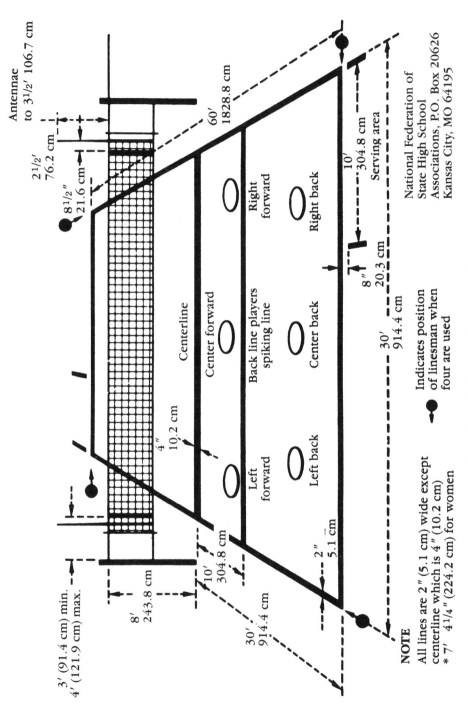

B-4 High school volleyball court markings.

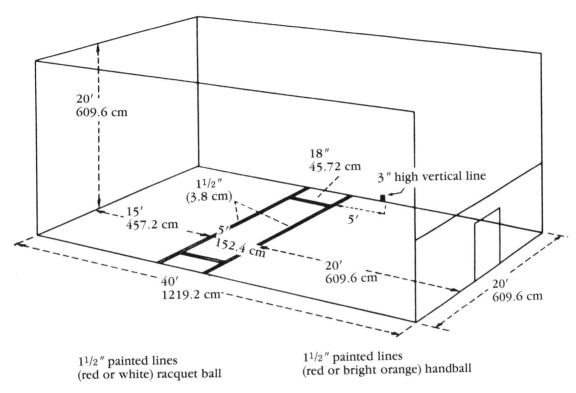

1¹/₂″ painted lines
(red or white) racquet ball

1¹/₂″ painted lines
(red or bright orange) handball

The United States Handball Association and United States Racquetball Association, 4101 Dempster Street, Skokie, IL 60076 (Joe Ardito-Racquetball) (Bob Peters-Handball)

RACQUETBALL FOUR-WALL RULES

Rule 2.1 - Court. The specifications for a standard four-wall racquetball court are:

(a) Dimensions. The dimensions shall be 20 feet high and 40 feet long with each back wall at least 12 feet high.

(b) Lines and Zones. Racquetball courts shall be divided and marked on the floors with 1¹/₂″ wide red or white lines as follows:

   (6) Receiving Lines. Five feet back of the short line, vertical lines shall be marked on each side wall extending 3 inches from the floor. The back edges of receiving lines shall be five feet from the back edge of the short line.

B-5 Four-wall handball and racquetball court markings.

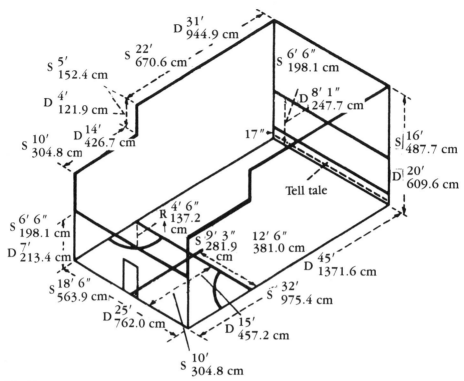

Markings are
inside dimensions
as lines are out of
bounds (except in
service areas).

1980 United States Squash Racquets Association
211 Ford Road, Bala-Cynwyd, PA 19004
(215) 667-4006 Mr. Kingsley, Executive Director

Dimensions shown above preceded by ''S'' are for signals court; by a ''D''
are for a doubles court.

Dimensions for red lines to be painted on floor and front wall are to center
of 1 '' (2.5 cm) wide line.

B-6   Standard squash court markings.

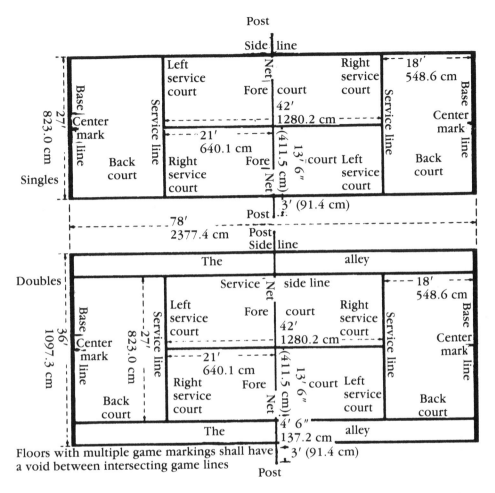

Floors with multiple game markings shall have a void between intersecting game lines

The space on each side of the net between the service-line and the side-line shall be divided into two equal parts called the service-courts by the center service-line, which must be 2″ (5.1 cm) in width, drawn half-way between and parallel with, the side-lines. Each base-line shall be bisected by an imaginary continuation of the center service-line to a line 4″ (10.2 cm) in length and 2″ (5.1 cm) in width called the center mark drawn inside the court, at right angles to and in contact with such base-lines. All other lines shall not be less than 1″ (2.5 cm) nor more than 2″ (5.1 cm) in width, except the base-line, which may be 4″ (10.2 cm) in width, and all measurements shall be made to the outside of the lines.

1980 Court Marking

United States Tennis Association, Education and Research Center, 729 Alexander Road, Princeton, NJ 08540
(609) 452-2580 Miles Dumont

B-7 Tennis court markings.

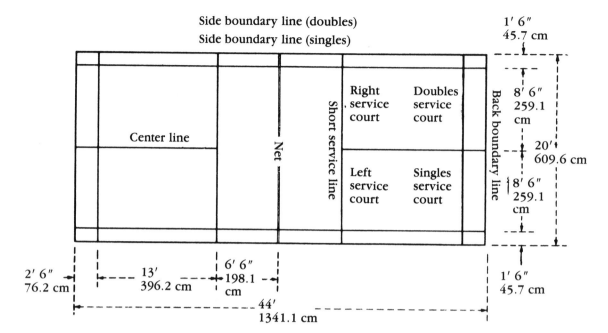

Side boundary line (doubles)
Side boundary line (singles)

Right service court

Doubles service court

Center line

Net

Short service line

Left service court

Singles service court

Back boundary line

1' 6"
45.7 cm

8' 6"
259.1 cm

20'
609.6 cm

8' 6"
259.1 cm

1' 6"
45.7 cm

2' 6"
76.2 cm

13'
396.2 cm

6' 6"
198.1 cm

44'
1341.1 cm

U.S. Badminton Association (ABA)
P.O. Box 237, Swartz Creek, MI 48473
(313) 655-4502   Mr. Eli

All lines to be 1 1/2" (38.1 mm) wide, preferably in white or yellow.

In marking the court, the width (1 1/2 inches) of the center lines shall be equally divided between the right and left service-courts; the width (1 1/2 inches each) of the short service line and the long service line shall fall within the 13-foot measurement given as the length of the service-court; and the width (1 1/2 inches each) of all other boundary lines shall fall within the measurements given.

B-8   Badminton court markings.

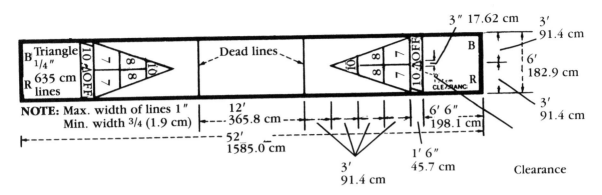

NOTE: Max. width of lines 1"
Min. width 3/4 (1.9 cm)

B-black | When playing doubles
R-red

Floors with multiple game markings shall have
a void between intersecting game lines

All line dimensions *must* be measured from line centers.

National Shuffleboard Association, Inc. President, Howard Hawkins
311-17th Avenue North, St.Petersburg, FL 33704, (813) 896-2240

B-9 Standard shuffleboard court markings.

Glossary

air-dried lumber   Lumber that has been piled in yards or sheds for any length of time. For the United States as a whole, the minimum moisture content of thoroughly air-dried lumber is 12 to 15%; the average is somewhat higher. In the South, the source of most American hardwoods, air-dried lumber may be no lower than 19%.

anchor bolts   Bolts to secure a wooden sill plate to a concrete or masonry wall or floor.

asphalt   Most native asphalt is a residue from evaporated petroleum. It is insoluble in water, but soluble in gasoline and it melts when heated. Used widely in building for waterproofing floors, flooring tile, roof covers, exterior walls, etc.

backband   A simple molding that is sometimes used around the outer edge of a plain rectangular casing as a decorative feature.

backfill   The replacement of excavated earth into a trench around and against a basement foundation.

baseboard   A board that is placed against a wall around a room next to the floor to give a proper finish between the floor and wall. Also called a base.

base molding   Molding used to trim the upper edge of interior baseboard.

base shoe   Molding used next to the floor on interior baseboards. Sometime called a carpet strip.

beam   A structural member that transversely supports a load.

bearing partition   A partition that supports any vertical load in addition to its own weight.

blind nailing   Nailing in such a way that the nail heads are not visible on the face of the work; usually done at the tongue of matched boards.

block flooring   Wood flooring made of blocks (squares or rectangular pieces of wood), rather than strips.

bodied linseed oil Linseed oil that has thickened in viscosity by suitable processing with heat or chemicals.

boiled linseed oil Linseed oil in which enough lead, manganese, or cobalt salts have been incorporated to make the oil harden more rapidly when spread in thin coatings.

brace An inclined piece of framing lumber that is applied to a wall or floor to stiffen the structure.

bridging Small wood or metal members that are inserted in a diagonal position between the floor joists at midspan to act both as tension and compression members for the purpose of bracing the joists and spreading the action of the loads.

butt joint The junction of the ends of two timbers or flooring boards in a square-cut joint.

condensation Beads or drops of water (and frequently frost in extremely cold weather) that accumulate on the inside of the exterior covering of a building when warm, moisture-laden air from the interior reaches a point where the temperature no longer permits the air to sustain the moisture it holds.

cove molding A molding with a concave face that is used as trim or to finish interior corners.

crawl space A shallow space between the living quarters of a basement-less house that is normally enclosed by the foundation wall.

cross bridging Diagonal bracing between adjacent floor joist spans to prevent the joists from twisting.

crown molding A molding used on a cornice or whenever an interior angle is to be covered.

cut-in-brace Nominal 2-inch thick members, usually 2 x 4s, that are cut in between each stud diagonally.

d—Penny As applied to nails, it originally indicated the price per hundred. The term now serves as a measure of nail length.

dado A rectangular groove across the width of a board or plank.

direct nailing Nailing perpendicular to the initial surface or the junction of the pieces joined. Also called face nailing.

drywall Interior covering material, such as gypsum board or plywood, which is applied in large sheets or panels.

drywall construction A type of construction in which the interior wall finish is applied in a dry condition, generally in the form of sheet materials or wood paneling, as contrasted to wet plaster.

fascia A flat board or face used sometimes by itself, but usually in combination with moldings, which are often located at the outer face of the cornice.

filler (wood) A heavily pigmented preparation used for filling and leveling off the pores in open-pored wood.

footing A masonry section, usually concrete, in a rectangular form that is wider than the bottom of the foundation wall or pier it supports.

foundation The supporting portion of a structure below the first floor construction, or below grade, that includes the footings.

frost line   The depth of frost penetration in soil. This depth varies in different parts of the country.

girder   A large principal beam of wood or steel that is used to support concentrated loads at isolated points along its length.

grain   The direction, size, arrangement, appearance, or quality of the fibers in wood.

grain, edge   Lumber that has been sawed parallel to the pith of the log and approximately at right angles to the growth rings, i.e., the rings form an angle of less than 45 degrees with the surface of the piece. Also called vertical grain.

grain, flat   Lumber that has been sawed parallel to the pith of the log and approximately tangent to the growth rings, i.e., the rings form an angle of less than 45 degrees with the surface of the piece.

hardwood   Wood from deciduous trees, such as maple, oak, and cherry that is known for its hard and compact substance.

header   A beam that is placed perpendicular to the joists and to which the joists are nailed in framing.

heartwood   The wood that extends from the pith to the sapwood, the cells of which no longer participate in the life process of the tree.

insulation   Any material high in resistance to heat transmission that when placed in the floors, walls, or ceiling of a structure will reduce the rate of heat flow.

joint   The space between the adjacent surfaces of two members or components that are joined and held together by nails, glue, cement, mortar, or other means.

joist   One of a series of parallel beams, usually 2 inches in thickness, that are used to support floor and ceiling loads and are supported in turn by larger beams, girders, or bearing walls.

kiln-dried lumber   Lumber that has been kiln-dried, often to a moisture content of 6 to 12%. Common varieties of softwood lumber, such as framing lumber, are dried to a somewhat higher moisture content.

knot   In lumber, the portion of a branch or limb of a tree that appears on the edge or face of a piece.

lumber   The wood product of the sawmill and the planing mill that is not further manufactured other than by sawing, resawing, and passing lengthwise through a standard planing machine, crosscutting to length and matching. Boards are lumber less than 2 inches thick and 2 or more inches wide. Dimension lumber is hard lumber from 2 to, but not including, 5 inches thick and 2 or more inches wide.

matched lumber   Lumber that is depressed and shaped on one edge in a grooved pattern and on the other edge in a tongued pattern.

millwork   Generally, all building materials made of finished wood and manufactured in millwork plants and planing mills, but not flooring, ceiling, or siding.

miter joint   The joint of two pieces at an angle that bisects the joining angle. For example, the miter joint at the corner of a floor (90 degrees) is made at a 45-degree angle.

molding   A wood strip that has a curved or projecting surface that is used for decorative purposes.

mortise   A slot cut into a board, plank, or timber, usually edgewise, to receive the tenon of another board, plank, or timer to form a joint.

nonbearing wall   A wall that supports no load other than its own weight.

notch   A crosswise rabbet at the end of a board.

O.C.   On center: the measurement of spacing for studs, rafters, joists, and the like in a building from the center of one member to the center of the next.

paper, building   A general term for paper, felts, and similar sheet materials that are used in buildings without reference to their properties or uses.

parquet   Inlaid mosaic of wood, used especially for floors. Technically, a floor of parquetry or wood mosaics in individual or block pieces is called a parquet floor.

pier   A column of masonry, usually rectangular in horizontal cross section, that is used to support other structural members, such as flooring joists.

pitch pocket   An opening that extends parallel to the annual rings of growth that usually contain or have contained either solid or liquid pitch.

pith   The small, soft core at the original center of a tree around which wood formation takes place.

plywood   A piece of wood that is made of three or more layers of veneer that are joined with glue and usually laid with the grain of adjoining plies at right angles. Almost always an odd number plies are used to provide balanced construction.

pores   Wood cells of comparatively large diameter that have open ends and are set one above the other to form continuous tubes. The openings of the vessels on the surface of a piece of wood are referred to as pores.

quarter round   A small molding that has a cross section in the shape of a quarter circle.

rabbet   A rectangular longitudinal groove that is cut in the corner edge of a board or plank.

sapwood   The outer zone of wood that is next to the bark. In the living tree, it contains some living cells, as well as dead and dying cells. In most species, it is a lighter color than the heartwood. In all species, it is lacking in decay resistance.

screed   A small strip of wood that is laid under the flooring or subflooring and above a concrete slab to separate the two.

sealer   A finishing material, either clear or pigmented, that is usually applied directly over uncoated wood to seal the surface.

sleeper   Usually a wood member that is embedded in concrete, as in a floor, that serves to support and to fasten subflooring or flooring.

softwood   Wood from conifer trees such as Douglas fir, hemlock, and southern pine.

soil cover A light covering of plastic film, roll roofing, or similar material that is used over the soil in crawl spaces to minimize moisture permeation of the area.

sole plate The bottom horizontal member of a frame wall.

solid bridging A solid member that is placed between adjacent floor joists near the center of the span to prevent the joists from twisting.

span The distance between structural supports such as walls, columns, piers, beams, girders, and trusses.

story The part of a building between any floor and the floor or roof above it.

stringer A timber or other support for cross members in floors or ceilings.

strip flooring Wood flooring that consists of narrow, matched strips, often of tongue-and-groove hardwood.

stud One of a series of slender wood or metal vertical structural members that are placed as supporting elements in walls and partitions.

subfloor Boards or plywood that are laid on joists over which a finish floor is to be laid.

termites Insects that superficially resemble ants in size, general appearance, and habit of living in colonies. Hence, they are frequently called white ants. Subterranean termites establish themselves in buildings, not by being carried in with lumber, but by entering from ground nests after the building has been constructed. If unmolested, they eat out the woodwork and leave a shell of wood to conceal their activities. Damage may proceed so far as to cause the collapse of parts of a structure before discovery. There are about 56 species of termites known in the United States, but the two major ones, classified by the manner in which they attack wood, are ground-inhabiting or subterranean termites (the most common) and dry-wood termites, which are found almost exclusively along the extreme southern border and the Gulf of Mexico in the United States.

termite shield A shield, usually of noncorrodible metal, that is placed in or on a foundation wall or other mass of masonry or around pipes to prevent the passage of termites.

toenailing To drive a nail at a slant with the initial surface in order to permit it to penetrate into a second member.

tongue-and-groove Boards or planks that are machined in such a manner that there is a groove on one edge and a corresponding tongue on the other. The most popular dressing joint for hardwood flooring. Also known as dressed and matched.

trim The finish materials in a building, such as baseboards or cornice moldings, that are applied around openings or at the floor and ceiling of rooms.

turpentine A volatile oil that is used as a thinner in paints and as a solvent for varnishes. Chemically, a mixture of terpenes.

underlayment A material that is placed under finish coverings, such as flooring or shingles, to provide a smooth, even surface for applying the finish.

vapor barrier   Material that is used to retard the movement of water vapor into floors and walls and to prevent condensation in them. It is usually considered as having a perm value of less than 1.0 and is applied separately over the warm side of exposed walls or as a part of batt or blanket insulation.

varnish   A thickened preparation of drying oil or drying oil and resin that is suitable for spreading on surfaces to form continuous, transparent coatings or for mixing with pigments to make enamels.

vehicle   The liquid portion of a finishing material; it consists of the binder and volatile thinners.

veneer   Thin sheets of wood that are made by rotary cutting or slicing a log. Hardwood veneers are often used for flooring.

volatile thinner   A liquid that evaporates readily and is used to thin or reduce the consistency of finishes without altering the relative volumes of pigments and nonvolatile vehicles.

wane   Bark, or the lack of wood from any cause, on the edge or corner of a piece of wood.

wood rays   Strips of cells that extend radially within a tree and vary in height from a few cells in some species to 4 inches or more in oak. The rays serve primarily to store food and to transport it horizontally in the tree.

Index